IRRIGATED CROPS
AND
THEIR MANAGEMENT

IRRIGATED CROPS
AND
THEIR MANAGEMENT

Roger Bailey

FARMING PRESS

First published 1990

Copyright © Roger Bailey, 1990

British Library Cataloguing in Publication Data
Bailey, Roger
 Irrigated crops and their management.
 1. Agricultural land. Irrigation. Economic aspects
 338.162

ISBN 0-85236-205-6

Published by Farming Press Books
4 Friars Courtyard, 30—32 Princes Street
Ipswich IP1 1RJ, United Kingdom

*Distributed in North America
by Diamond Farm Enterprises,
Box 537, Alexandria Bay, NY 13607, USA*

Cover design by Andrew Thistlethwaite

Phototypeset by Galleon Photosetting, Ipswich
Printed and bound in Great Britain by Biddles Ltd
Guildford and King's Lynn

Contents

Colour Plates

Figures

Preface

There are many books covering the techniques of irrigation in the drier parts of the world, but there is very little published information that is specifically aimed at our situation in the United Kingdom and in north-west Europe. For some time I have recognised the need for a book detailing the economics of irrigation, the water requirements of various crops and the planning of irrigation application in our climate. This book is an attempt to collect such information together in the hope that it will be of use to farmers, advisers and students with an interest in irrigation.

In preparing this book, I have been helped by numerous friends and colleagues in ADAS and the Agriculture and Food Research Council (AFRC) Institutes. I wish to acknowledge with grateful thanks their considerable assistance in providing material and advice. In particular, I would like to thank Bob Hart and John Martindale for their invaluable critical comment on the text.

Finally, my thanks to Maire, Glennyz and Glyn for typing the manuscript.

August 1989 ROGER BAILEY

IRRIGATED CROPS
AND
THEIR MANAGEMENT

Chapter 1

Irrigation Planning

Many crops grown in lowland areas of north-western Europe regularly experience moisture stress, and as a result their growth is often limited. In an increasingly competitive situation requiring premium quality as well as high yields, a reliable and continuous supply of water is an essential part of success in growing many high-value crops.

In 1982, farmers in England and Wales irrigated 100,000 ha; by 1984 this area had increased to 140,000 ha. Similarly, irrigation is important in other north-western European countries. In 1984 there were 90,000 ha irrigated in Norway, 250,000 in West Germany, 100,000 in Finland, 260,000 in Holland, 160,000 in Sweden and a staggering 400,000 ha in Denmark, all northern European countries with climates superficially less demanding of irrigation than the UK.

Unlike the hotter, arid parts of the world where irrigation is required in large quantities in order that crops may survive, our requirements in the UK and north-western Europe are moderate. We know from experience that crops usually survive without irrigation and, indeed, the majority of farmers continue to produce high yields without irrigation. In certain areas, however, and on certain soils, we irrigate in order to obtain higher yields, better quality and a reliable continuity of supply.

Several years ago, I toured the United States of America in order to study irrigation practices. I was repeatedly asked the same questions by research workers over there. As irrigation has such relatively marginal benefits in a north-west European climate, why do we make such a painstaking effort to plan it accurately? Why do we not just crudely apply some water during dry spells and leave it at that? My answers were based on straightforward economics. It is precisely because our returns from irrigation are so marginal that we have to apply water accurately. If we apply too little, the profits that we need to justify the cost will fail to materialise. If we apply too much, the cost will be too great to make the returns worthwhile, and also we could suffer yield losses as a result of nutrient leaching. It is only by planning and applying irrigation accurately that we can economically produce the high yields and good quality that we are seeking.

1

The objective of irrigation planning is two-fold. First, it is important to match investment to the amount of irrigation that is economically worthwhile on each individual farm. In wetter areas, and on high moisture retentive soils, irrigation is not a good investment. Conversely, in dry areas, and where soils have a low water-holding capacity, irrigation can be of significant benefit. For some crops (e.g. iceberg lettuce), irrigation is essential if a high-quality product is to be produced consistently. Good planning involves the identification of those situations where irrigation is financially worthwhile, and estimating the likely quantities that are required to optimise the returns from investment.

The second objective is to ensure that the available irrigation water is applied at the correct time, and in a sufficient quantity to satisfy the demands of the crop, but not in excess of this requirement. This is done by continually monitoring the moisture status of the soil, and applying water before the soil dries to the critical state at which crop yield or quality are affected.

The techniques used to plan irrigation correctly have been the subject of much debate. In the rest of this book I shall attempt to place these techniques in perspective, and describe how each farmer can assess his need for irrigation. I shall also describe how to judge when irrigation is required and, just as importantly, how irrigation interacts with other aspects of good crop husbandry.

An Introduction to Soil—Plant—Water Relationships

Water is of prime importance to all living organisms; the plants that man has developed and cultivated as agricultural crops are no exception to this rule. This need is fairly easy to satisfy for aquatic plants, which are surrounded by water. In the terrestrial environment, however, water relations are more complex. The plant can only obtain water from a localised part of its environment, namely, the soil or other rooting medium. At the same time, the upper parts of the plant are continually subjected to drying conditions. Terrestrial plant groups, including our cultivated crops, have had to develop mechanisms to overcome these problems. To obtain water efficiently from the soil, terrestrial plants have developed complex root systems, involving a considerable degree of branching. The total root length of a single wheat plant is estimated to measure over forty miles. To prevent desiccation of the upper parts, most terrestrial plants are covered in a waterproof layer, which helps to retain the water they extract from the soil.

The raw materials required by green plants to build the substances they need for growth and energy are water, carbon dioxide and a variety of essential elements, some in trace amounts. Carbon dioxide enters as gas from the atmosphere, but minerals enter the plant in solution via the roots. The outer waterproof covering of the leaves is perforated with tiny adjustable holes through which carbon dioxide enters, and oxygen and water vapour can exit. These perforations are known as *stomata*. The density of stomata varies with the species, but a common figure quoted for many plants is 300 per square millimetre. While the stomata are open, carbon dioxide passes through into the plant. At the same time, water carrying essential nutrients in solution passes into the roots, moves upward through the plant depositing the nutrients where they are required, evaporates within the leaf and passes through the open stomata into the atmosphere as water vapour. This process is called *transpiration*. The water lost from the leaves is then replaced by more water, carrying more of the essential nutrients from the soil, via the roots. There is thus a continuous movement of water from the soil to the atmosphere, called the *transpiration stream.*

The amount of water used by plants in this way is considerable, and a dense crop can transpire as much as 250 tonnes of water per hectare per week in June/July in England. This is an inconvenient measure of water usage, however. It is usual practice to describe rainfall in terms of depth (e.g. 10 mm rain), and water usage by crops is best described in similar terms. Thus 250 tonnes of water per hectare is more meaningful if expressed as 25 mm water.

Some water is also lost from the soil by direct surface evaporation. When calculating water loss from the soil to the atmosphere, it is convenient to consider transpiration and evaporation together, and the resulting water loss is then referred to as *evapotranspiration.*

This whole process runs smoothly as long as there is a plentiful supply of moisture within the soil, allowing the plant to take up water at the same rate as it is being lost by evaporation through the stomata, maintaining a balance within the plant. However, if the moisture supply in the soil becomes limiting, the rate of uptake is restricted, and the plant will become depleted of water and ultimately wilt unless it can regulate the outflow through the stomata. Regulation is achieved simply by closing the stomata, but this also has the effect of restricting inflow of carbon dioxide into the plant. The net result is a reduced rate of growth, lower dry matter accumulation and, usually, lower yields.

The objective of irrigation is to keep the moisture supply in the soil at a sufficient level to prevent the stomata closing, but to avoid an over-supply of water which is costly and adds to the risk of nutrient leaching. If we are to achieve this optimum course, it is necessary to understand precisely when the moisture supply from the soil becomes limiting.

FIELD CAPACITY AND SOIL MOISTURE DEFICIT (SMD)

After heavy rain, a free-draining soil is temporarily *saturated*, and all the pores of the soil are filled with water (see Figure 1.1(*a*)). This is usually a short-lived situation, as the surplus water soon starts to drain downwards until the soil is holding the maximum quantity of water that it can retain against the pull of gravity. The force holding this water is *surface tension*, and a soil in this state is said to be at *field capacity* (see Figure 1.1(*b*)). It is convenient to consider this water in terms of the suction required to remove some of it. At field capacity, this is equal to 0.05 bar, and we say that the soil moisture tension

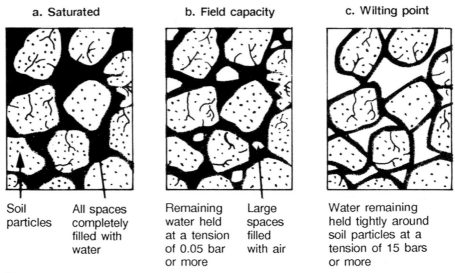

a. Saturated		b. Field capacity		c. Wilting point
Soil particles	All spaces completely filled with water	Remaining water held at a tension of 0.05 bar or more	Large spaces filled with air	Water remaining held tightly around soil particles at a tension of 15 bars or more

Figure 1.1   Diagrammatic representation of soil at various stages.

is 0.05 bar. However, it is uncommon to find all layers of a soil profile at field capacity at the same time. The process of drainage from saturation to field capacity is not instantaneous, but takes place over a period of days, or even weeks, depending on soil type. The water drains from the upper layers initially, and so, at any one instant, there will be a layer of soil that is at field capacity but lower layers will still be saturated. Also, upper layers soon become drier than field capacity, because evaporation from the surface occurs while the lower layers are still draining. Nevertheless, for most practical purposes, we can assume that light soils drain to field capacity throughout the rooting zone within a couple of days after being saturated.

As a crop removes water from the soil, the soil is described as having a certain *soil moisture deficit (SMD)*. The SMD at field capacity is zero. If a crop extracts 50 mm of water from a soil, and there has been no rainfall or irrigation to replenish it, the soil is said to have an SMD of 50 mm. If, during the following few days, the crop extracts a further 10 mm of water, the SMD is then 60 mm. If rainfall then adds 20 mm of water back to the soil, the SMD becomes 40 mm. As this process occurs, and water is removed from the soil, the water remaining is that which is held at a greater tension, i.e. it requires a greater suction to remove it.

CRITICAL SOIL MOISTURE DEFICITS

I have previously referred to the process of water passing out through the open stomata and evaporating into the atmosphere as *transpiration*. The rate at which this occurs is known as the *actual transpiration rate*. The driving force which determines this is the weather, but it is also regulated by the type of crop and the availability of moisture in the soil. It is increased by bright sunshine, high temperatures or strong winds. It is decreased in cool, cloudy and still conditions, and also if there is a significant restriction on the availability of soil moisture, preventing the crop from extracting water at the required rate. When there is no restriction on soil moisture, and the rate for a given crop is determined solely by the weather, it is said to be proceeding at its full potential which is referred to as the *potential transpiration rate*.

At field capacity, or when the SMD is low, water supply is unrestricted and the actual transpiration rate is equal to the potential transpiration rate. As the SMD increases, water becomes more difficult to extract, and plants respond by closing stomata and thereby reducing water loss to the atmosphere. When this happens, the actual transpiration rate falls below the potential rate. This is an important stage, and the SMD at which it occurs is known as the *critical SMD*, because this closing of the stomata will also result in less intake of carbon dioxide, and therefore lower rates of photosynthesis and potentially lower yields.

If the soil dries beyond the critical SMD, plants continue to extract water, but at a rate that is continually reducing. Eventually, the only remaining water in the soil is that which is held more strongly than the suction exerted by plant roots, and the plants can extract no more. At this point, transpiration ceases, and the plants wilt and die. The soil is then said to be at *permanent wilting point*. This is usually considered to be the stage when the remaining water is being held by the soil at a tension of 15 bars (see Figure 1.1(*c*)). Thus, at any given time, the weather determines the potential rate of transpiration, but the actual rate is dependent on both the weather and the availability of soil moisture.

In most situations, our objective with irrigation is to keep the SMD below the critical level. This is done by triggering irrigation whenever the SMD approaches the critical SMD. For this reason some authors have used the term *trigger SMD* instead of critical SMD. I prefer to distinguish between the two, however. The critical SMD is an estimable value dependent on soil, crop and weather conditions; the trigger SMD is merely an arbitrary point chosen by the farmer or grower to trigger his irrigation. The two are usually the same, but not always.

Consider a crop of very low value in a situation where irrigation is very expensive. The farmer may allow the soil to dry way beyond the critical SMD, and only irrigate at the final moment to prevent the crop actually dying. In this case his chosen trigger SMD would be a much higher value than the critical SMD.

AVAILABLE WATER CAPACITY

From the above discussion, it is apparent that not all the water in a soil is available to a crop, but only that part of the water held at a tension of less than 15 bars. This water is referred to as the *available water*. Soils are frequently described in terms of their *available water capacity* (AWC), which is expressed as a percentage. Thus, when at field capacity, a soil with an AWC of 17 per cent holds available water equivalent to 17 per cent of its volume.

Different soils have widely differing AWCs, depending on their texture and structure. Silt loams, for example, have much more water available for plants to extract than do sands (see Figure 1.2 for a diagrammatic explanation). Silt loams have a high AWC, sands a low AWC.

While an AWC tells us something about the potential of the soil, it still does not tell us how much water is actually available to a particular crop, as it takes no account of depth of rooting. A crop with deep roots will have more water available to it than a shallow rooting crop. When planning irrigation, it is useful to describe the amount of available water found within the rooting zone of the crop in question. Throughout the rest of this book, the term *root zone capacity* will be used to describe this water (see Figure 1.3). Thus a soil with an AWC of 17 per cent will have 170 mm root zone capacity for a crop that rooted to 1 metre, but only 85 mm root zone capacity for a crop with a rooting depth limited to 50 cm.

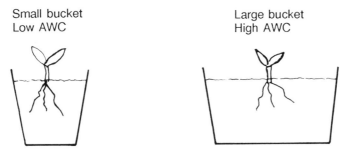

Figure 1.2   Effect of soil type upon available water. Plants grown in large buckets have more water available than those in small buckets. Plants in high AWC soils have more water available than those in low AWC soils.

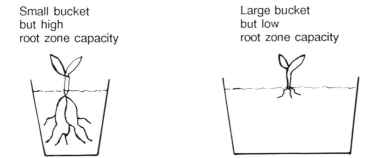

Figure 1.3   Effect of rooting depth upon available water. A shallow rooting system has little water available, even in a high AWC soil.

A major purpose of this chapter has been to explain the jargon used in irrigation. This is important because the fundamental characteristics of plant—soil—water relationships measured in these terms must be understood for any particular farm on which a reasoned financial appraisal is to be made for irrigation, or if irrigation scheduling is to be planned correctly. The first steps for either exercise are:

Stage 1   Examine the soil, using auger, spade, soil pit and tape measure.
Stage 2   Determine the AWC.
Stage 3   Estimate the root zone capacity for each crop.
Stage 4   Estimate the critical SMD for each crop.

These steps are described in detail in chapter 2. Subsequent chapters will describe the estimation of the irrigation requirement and the expected benefits with each crop, along with day-to-day irrigation planning.

REFERENCES AND FURTHER READING

ASLYNG, H. C. (1984), 'Soil water capacity, climate and plant production. The development and future of irrigation', *Proceedings of the North-Western European Irrigation Conference*, Billund, Denmark, 2—39.

BERG, E. (1984), 'The role of irrigation in Norway', *Proceedings of the North-Western European Irrigation Conference*, Billund, Denmark, 49—57.

BOHEEMEN, P. J. M. VAN (1984), 'Supplement irrigation in the Netherlands', *Proceedings of the North-Western European Irrigation Conference*, Billund, Denmark, 89—98.

ELONEN, P. (1984), 'The role of irrigation in Finland', *Proceedings of the North-Western European Irrigation Conference*, Billund, Denmark, 76—84.

FEYEN, J. (1984), 'The role of irrigation in Belgium', *Proceedings of the North-Western European Irrigation Conference*, Billund, Denmark, 85—8.

HART, R. (1984), 'The role of irrigation in England and Wales', *Proceedings of the North-Western European Irrigation Conference*, Billund, Denmark, 40—8.

LINNER, H. (1984), 'Irrigation in Sweden', *Proceedings of the North-Western European Irrigation Conference*, Billund, Denmark, 99—104.

MAFF (1982) 'Irrigation', *Reference Book 138*.

MAFF, 'Agricultural Statistics United Kingdom. Irrigation of Outdoor Crops' (various dates).

WOLFF, P. (1984), 'The role of irrigation in the Federal Republic of Germany', *Proceedings of the North-Western European Irrigation Conference*, Billund, Denmark, 58—75.

Chapter 2

Calculation of Critical Soil Moisture Deficits

In the previous chapter the concept of critical soil moisture deficit (SMD) was explained. It is an important stage because, if the soil dries beyond the critical SMD, plants will extract water at a reduced rate with a consequent penalty in yield or quality.

In reality, however, the critical SMD is not fixed, but varies according to weather conditions. There have been several studies which showed that the weather on a particular day had a greater effect than soil moisture status in determining whether plants were experiencing moisture stress. On a hot, bright day with a high transpiration demand, a crop is more likely to be experiencing stress than it is on a dull, cloudy day.

Stress is caused when there is an imbalance between transpiration demand and water supply to the crop. When the transpiration demand is low, crops may be able to meet this demand adequately even though soil moisture reserves are low, i.e. not until the deficit is very high does it become 'critical'. Conversely, when the transpiration demand is high, plants may experience stress even at low deficits. The level at which the deficit becomes 'critical' varies with the transpiration demand.

It follows that the combination of conditions with the greatest detrimental effect on plant growth is that of high transpiration demand along with low moisture reserves. We try to prevent this combination by irrigating. We cannot control transpiration demand (except possibly with misting: see chapter 9), but we can control soil moisture reserves with irrigation.

Unfortunately, we cannot forecast transpiration demand far in advance, and this leaves us with a practical problem. Do we keep the soil moisture status at levels sufficient only for conditions of average transpiration demand, or do we cater for conditions of higher demand? The first strategy is less costly but will result in considerable drought losses if the weather becomes hot and dry. The second strategy is more expensive but will avoid most losses in a hot, dry year. With high-value crops, it is usual to adopt this second strategy, accepting that it will inevitably involve high amounts of irrigation, which may prove to have been unnecessary if the season remains cool and cloudy.

The critical deficits referred to in most publications, including this one, are those that operate in conditions of high transpiration demand, unless otherwise stated.

How is the critical SMD determined for a crop? The various methods will be described in this chapter.

MEASURING CRITICAL SMDs BY EXPERIMENT

Critical SMDs have been determined by experiment for a range of crops. An experiment will typically consist of a series of treatments, each of which trigger irrigation at a different SMD, e.g.: (*a*) irrigate every time SMD reaches 25 mm; (*b*) irrigate every time SMD reaches 35 mm; (*c*) irrigate every time SMD reaches 45 mm, etc. Unfortunately, such experiments tend to be very inefficient. Natural rainfall often intervenes and prevents the higher trigger SMDs being achieved more than once or twice, if at all, in a season, unless a mobile shelter is used to cover the plots during rainfall (see Colour plate 1). A further problem is that, as already stated, critical SMDs are not fixed, but vary according to weather conditions. This type of experiment provides a clear indication of critical SMDs only when conducted in a hot, dry year. As a result, it is necessary to keep repeating the experiment until there is enough data obtained over a number of years to pick out confidently the highest deficit that can be allowed without loss of yield or quality. The results from a series of such trials on maincrop potatoes at Gleadthorpe EHF (Experimental Husbandry Farm) are shown in Table 2.1.

Table 2.1   Irrigation regimes and ware yield (t/ha) of maincrop potatoes at Gleadthorpe EHF, 1959–70

| | *Irrigation applied whenever SMD reached the following:* | | | | | | | | *None applied* |
	10 mm	*20 mm*	*25 mm*	*30 mm*	*35 mm*	*40 mm*	*50 mm*	*55 mm*	
1959	—	—	34.3	—	—	—	29.0	—	8.8
1960	36.3	38.3	—	36.5	—	—	—	—	29.3
1961	32.0	31.3	—	31.8	—	—	—	—	28.0
1964	—	38.0	—	—	—	34.8	—	—	32.0
1969	—	42.8	—	—	36.0	—	—	40.0	28.5
1970	—	48.0	—	—	47.8	—	—	45.8	27.3

Source: Gleadthorpe EHF, ADAS.

The difficulty of deciding on the critical SMD is immediately apparent. However, by the end of the 1964 experiment, the researchers felt able to conclude that the critical deficit was probably around 35 mm at this site. Experiments of this nature were repeated on another variety in 1981–3 and the results, shown in Table 2.2, are in agreement with the earlier work. In 1981, very high-yield responses were obtained from irrigation, and applying water at an SMD of 35 mm was as productive as irrigating at lower SMDs. In 1983, this treatment again worked well but yields showed a tendency to fall if the deficit was allowed to build up to 45 or 55 mm.

Table 2.2 Irrigation regimes and ware yield (t/ha) of maincrop potatoes at Gleadthorpe EHF, 1981–3

| | *Irrigation applied whenever SMD reached the following:* | | | | | |
	*Scab control**	*25 mm*	*35 mm*	*45 mm*	*55 mm*	*None applied*
1981	51.8	52.6	52.7	—	—	27.8
1982	41.5	—	40.8	39.8	—	32.6
1983	45.7	—	45.7	43.3	41.2	—

Source: Gleadthorpe EHF, ADAS.
Note: * Intensive irrigation, i.e. deficits kept below 18 mm for 6 weeks from tuber initiation and thereafter below 35 mm.

A second type of experiment involves regular monitoring of irrigated and unirrigated treatments for yield, canopy development and water use. Early in the season, before the unirrigated crop has experienced stress, the growth analysis measurements should be similar for both treatments. However, as the soil progressively dries on the unirrigated plots, they will start to show reduced growth compared with the irrigated ones. If the growth analysis is sensitive, it should be possible to pinpoint precisely when this difference in growth first occurred, and this point will correspond to the critical SMD. This approach also has its problems, however.

As an example, consider a crop such as peas, where we are only concerned with the final yield of a particular fraction, i.e. the seed. A moderate restriction in moisture supply early in the season will reduce haulm growth, but will not necessarily reduce the final yield of the peas. Thus the growth parameter being carefully measured only indicates a critical deficit if it is a parameter that is linked to final yield. This experimental approach is a useful one, but it is important to distinguish between a deficit that may be critical for total growth, and a deficit that is critical to the final yield of the harvested fraction.

One of the basic facts that everybody in farming appreciates is that different soils have different moisture characteristics. Crops grown in sandy soils generally suffer more from drought than those growing on heavier soils. This is because heavier soils have more water available for the crop, i.e. they have a higher available water capacity (AWC). We therefore need to know the critical SMD for each crop in a wide range of soils.

Both types of experiment described here are used to determine critical SMDs, but both involve much detailed work and often have to be continued for many years before firm conclusions can be drawn. It is not, therefore, possible to carry out this research over the whole range of soil textures found within UK and north-west European agriculture. The experiments are conducted at the various research stations in order to establish and test irrigation principles. For practical irrigation purposes we also need a simple technique of determining the critical SMD for each crop over the complete range of soil types. The technique suggested here involves calculating the amount of water available to a crop, taking soil texture and rooting depth into account. This is known as the *root zone capacity*. The critical deficit can then be estimated from the root zone capacity:

Critical SMD = root zone capacity × allowable depletion fraction

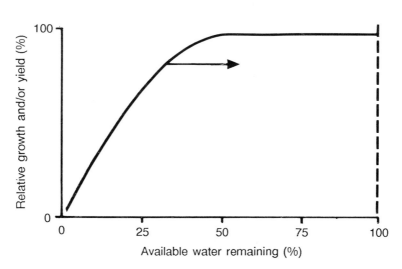

Figure 2.1   *Range of root zone available water that can be utilised before growth and/or yield is reduced. Note that the curve translates to the right with high transpiration rates, and with sensitive crops at certain growth stages.*   (Source: Stegman, 1983.)

The allowable depletion fraction is the proportion of the root zone capacity that can be safely depleted between irrigations without the crop becoming stressed (see Figure 2.1). Some authors have suggested that a fraction of one-half represents an average 'safe' level for a wide array of crops and soils. However, Doorenbos and Kassam described the dependency of this fraction upon the specific crop and the level of transpiration demand. A further discussion of this is presented later in this chapter.

In order to estimate a critical SMD for each situation, it is necessary to have a clear understanding of:

1. The effect of soil texture.
2. The rooting pattern of each crop and the calculation of root zone capacity at relevant points in the growing season.
3. The relationship between root zone capacity and critical SMD.

Effect of Soil Texture on Available Water

The profile of a soil can only be determined by digging and examining the side of a pit. We also need to know the texture of both the topsoil and the subsoil. In cases where there are more than two clear layers in the profile, a description of each will be required. Soil texture is usually determined by the simple procedure of rubbing the damp soil between finger and thumb -- but this requires a certain amount of training, and farmers often employ an adviser to assess this soil characteristic. Where very precise descriptions are required, soil texture can be assessed in the laboratory using particle size analysis, but this is not usually necessary for the purpose of irrigation planning. Once the soil texture has been established, the next step is to determine the available water capacity of the profile.

It is possible to measure the available water capacity (AWC) of a soil by taking soil cores to the laboratory, applying suction pressure to them and measuring the amount of water that is released. It is generally agreed that plants can extract moisture up to a tension of 15 bar, so the amount of water that can be removed between field capacity and 15 bar suction pressure is equivalent to the available water capacity. This method of determining AWC is, of course, very time-consuming and while some growers are prepared to pay a laboratory for this analysis, the majority require an easier and less costly technique.

The Soil Survey and Land Research Centre (formerly the Soil Survey of England and Wales) has used the laboratory technique to measure available water in over 1,000 soil profiles in England and Wales. In this study, it has been established that the major variation in AWC is due to soil texture, soil structural condition and whether a soil layer is a topsoil or a subsoil. Although plants can extract water to a tension of 15 bars, it is much easier for them to extract water that is held at 2 bars or less. This smaller fraction is referred to as the 'easily available water' and is also of use to us in irrigation planning. The results from the Soil Survey dataset have been summarised by Hollis of the Soil Survey and are presented in Table 2.3

Table 2.3   Estimation of available water up to 15 bars (and easily available water up to 2 bars) from soils of varying texture and structure

| | | Subsoil available water % (easily available %) | | |
Texture class	Topsoil available water % (easily available %)	Good structural conditions	Average structural conditions	Poor structural conditions
Clay	17 (10)	21 (15)	16 (8)	13 (7)
Silty clay	17 (10)	21 (15)	15 (8)	12 (7)
Sandy clay	17 (11)	19 (14)	15 (10)	13 (8)
Sandy clay loam	17 (11)	19 (14)	15 (10)	13 (8)
Clay loam	18 (11)	21 (14)	16 (10)	12 (7)
Silty clay loam	19 (10)	21 (12)	17 (10)	12 (6)
Silt loam	23 (15)	23 (17)	22 (14)	15 (9)
Fine sandy silt loam	22 (14)	22 (16)	21 (15)	15 (9)
Medium sandy silt loam	19 (11)	19 (13)	17 (11)	15 (9)
Coarse sandy silt loam	19 (11)	23 (17)	19 (11)	15 (7)
Fine sandy loam	18 (13)	22 (17)	18 (13)	17 (11)
Medium sandy loam	17 (11)	17 (13)	15 (11)	11 (8)
Coarse sandy loam	17 (11)	22 (15)	16 (11)	11 (8)
Loamy fine sand	18 (14)	15 (13)	15 (13)	—
Loamy medium sand	13 (9)	12 (9)	9 (6)	—
Loamy coarse sand	11 (7)	11 (7)	8 (6)	—
Fine sand	—	14 (12)	14 (12)	—
Medium sand	12 (8)	7 (5)	7 (5)	—
Coarse sand	—	5 (4)	7 (5)	—
Marine light silts*	—	33 (30)	28 (22)	—

Source: Soil Survey and Land Research Centre.
Notes: * Use these figures *only for subsoils* in marine alluvium where textures are fine sandy silt loam, fine sandy loam or loamy fine sand *and* most of the sand is finer than 0.1 mm.
   — Rare occurrences for which there are no data.

Reeve of the Soil Survey has kindly produced a key to interpretation of good, average and poor structural condition of subsoils and this is presented in Table 2.4.

Table 2.4   Guide for assessment of structural condition of soil

Structural condition	Field Properties
Good	Low density, high organic matter. Topsoils and upper subsoils under or recently under permanent grassland. Very well structured topsoils and subsoils in light alluvium.
Average	Medium density, average organic matter. Topsoils under regular arable use. Subsoils in alluvium or over chalk, limestone, sandstone, siltstone or light glacial drift. Unmottled upper subsoils above impermeable clays and heavy drifts.
Poor	High density, low organic matter. Badly compacted topsoils and ploughpans. Mottled subsoils in impermeable clays and heavy drifts.

Source: Soil Survey and Land Research Centre.

Where stones are present in a soil, they have to be taken into account in the calculation of available water. If there are less than 50 per cent stones by volume, then the calculation is straightforward. For example, where there are 20 per cent in a layer, then the calculated AWC for that layer is reduced by 20 per cent. In the rare situations where the proportion of stones exceeds 50 per cent, this simple correction will not be precise enough and it is necessary to take into account the AWC of the actual rock material, e.g. chalk, gravel, sandstone, etc.

Where a solid impenetrable layer of rock material exists, then soil available water is estimated only from those layers above the rock. However, where the rock material is mixed with soil, the amount of available water can be estimated. Values for different types of rock material have been suggested by the Soil Survey and are presented in Table 2.5. These figures are based on very few measurements and should be regarded as tentative, and only used where actual AWC measurements are unavailable.

The calculation of AWC is as follows:

$$AWC = \frac{(AWC \text{ of fine earth} \times \% \text{ fine earth}) + (AWC \text{ of stones} \times \% \text{ stones})}{100}$$

From Tables 2.3, 2.4 and 2.5 it is a simple procedure to estimate available water and easily available water from any particular soil profile. But, to calculate the amount of water available to a crop, it is also necessary to consider the rooting patterns of different crops.

Table 2.5 Available water capacity in stones and rock

Rock, gravel or stone type	Available water %	Easily available water %
All hard rocks or stones	1.0	0.5
Soft, medium or coarse grained sandstones	3.0	2.0
Soft, fine grained sandstones	5.0	3.0
Soft limestone	4.0	3.0
Chalk or chalk stones	10.0	7.0
Gravel with hard stones	2.0	1.0
Gravel with soft stones	5.0	3.0

Source: Soil Survey of England and Wales.

ROOTING PATTERNS AND THE CALCULATION OF ROOT ZONE CAPACITY

Studies of crop roots show that each crop has a particular rooting pattern. For example, potato roots do not grow deeply and 70 cm is a typical limit. It is therefore assumed that potatoes can extract the available water up to a tension of 15 bar in the upper 70 cm of the soil profile, and this amount represents the root zone capacity of the crop. From Table 2.3, the root zone capacity of potatoes can be calculated for a soil of any description.

Studies have also shown that many crops exhibit a more marked concentration of roots within the upper horizons, and a progressively decreasing amount with greater depth. Cereals are a good example, with most roots found in the upper 50 cm, but under favourable conditions some roots penetrate down to 1.2 metres or even more.

In order to take this type of rooting pattern into account, Hall *et al.* of the Soil Survey have suggested that the abstraction of water approaches 15 bar suction only where the highest concentration of roots is found. In the lower horizons, where root growth is limited, the crop can only extract the available water in the immediate vicinity of each root, i.e. water is extracted from localised pockets and some of the water in the soil remains unexploited. The proportion is difficult to estimate precisely and will depend on the specific concentration of roots in these layers, but Hall has suggested a simple technique that is sufficient for all practical purposes. In these lower horizons of limited rooting, the amount of water available to the crop is equivalent to the easily available water held to a tension of 2 bar (given in brackets in Table 2.3). This would appear to be a reasonable assumption and I have adopted it here.

For some crops, there are published studies that show how much root material exists at certain depths. Where I have referred to this information, I have arbitrarily chosen 0.3–0.5 cm root per cubic cm of soil as being the lowest density able to extract water to 15 bar tension. I have also chosen 0.01 cm root per cubic cm of soil as being the lowest density of root able to extract moisture at 2 bar tension. This is not based on conclusive measurement, but was chosen after discussion with colleagues (both within and outside the Agricultural Development and Advisory Service) involved in irrigation research. For some crops, quantitative measurements of rooting are unavailable, and for these the information presented here is based on loose descriptions in the literature such as 'most of the roots were found in the upper — cm, but some extended to — cm'.

Some crops, such as sugar beet, possess a root system that changes markedly throughout the irrigation season. The depth of rooting in August and September is over twice that in June. For such crops, it is insufficient to use one set of figures through the season; adjustment is necessary at intervals. Of course, most of our annual crops demonstrate some degree of root extension throughout the irrigation period, and it may be that this principle of adjusting the critical deficit should be used for them all. On the basis of the experimental evidence currently available, it appears that good results can be achieved for crops such as potatoes and cereals without resorting to these adjustments. There is much research work still required, however, and our techniques may become further refined as more evidence becomes available.

The rooting patterns recommended for irrigation planning are shown in Table 2.6. With the assistance of this rooting data, it is now possible to estimate the proportion of the available water in the soil profile that is actually accessible to a crop, i.e. the root zone capacity.

Root zone capacity water can be calculated simply by referring to Tables 2.3 and 2.6, as demonstrated in the following examples.

Example 1 (see diagram on page 19)

Consider a crop of maincrop potatoes grown in a field where the soil profile is 350 mm of loamy medium sand over a medium sand subsoil. From Table 2.6, the rooting pattern of maincrop potatoes is such that the crop can extract all the available water to a depth of 700 mm. From Table 2.3, the topsoil has an AWC of 13 per cent. As the depth of topsoil in this case is 350 mm, 13 per cent represents 45.5 mm of water. From the table, the medium sand subsoil has an AWC of 7 per cent.

Table 2.6 Rooting patterns of various crops as used in irrigation planning

Crop	Depth in cm of intensive rooting (capable of extracting available water, i.e. up to 15 bar)	Depth in cm of sparse rooting (capable of extracting easily available water only, i.e. up to 2 bar)	Source of information
Early potatoes	50	—	1
Maincrop potatoes	70	—	2
Sugar beet in June	25	50	3
Sugar beet in July	50	100	3
Sugar beet in August	80	130	3
Sugar beet in September	100	160	3
Winter cereals	80	120	2,4
Spring cereals	50	120	2
Oilseed rape	100	—	1
Spring field beans	40	70	5
Winter field beans	40	90	5
Peas	25	70	5,6
Broad beans	40	70	5
Onions	25	60	7
Turnips 6 weeks after emergence	10	40	6
Turnips 9 weeks after emergence	65	80	6
Turnips 13 weeks after emergence	80	—	6
Carrots	75	—	1
Cauliflower 5 weeks after emergence	5	25	6
Cauliflower 9 weeks after emergence	15	70	6
Cauliflower 11 weeks after emergence	30	70	6
Cauliflower 14 weeks after emergence	50	80	6
Lettuce 6 weeks after emergence	5	30	6
Lettuce 8 weeks after emergence	15	60	6
Strawberries	45	60	8
Deciduous orchards	100—200	—	9

Sources include:
1 Aslyng (1984); 2 Hall *et al.* (1977); 3 Dunham (1987). Dunham (personal communication); 4 Barraclough and Leigh (1984). Barraclough (personal communication); 5 Hebblethwaite (1982). Heath and Hebblethwaite (1985). Hebblethwaite (personal communication); 6 Greenwood *et al.* (1982); 7 Brewster (1977); 8 Hughes (1965). Higgs and Jones (1989); 9 Doorenbos and Pruitt (1977).

The crop only roots to 700 mm, so there is 350 mm of subsoil within rooting range. The 7 per cent AWC in the subsoil is thus 24.5 mm. Adding the two amounts together, the total water available to the crop of potatoes in this soil (the root zone capacity water) is 70 mm.

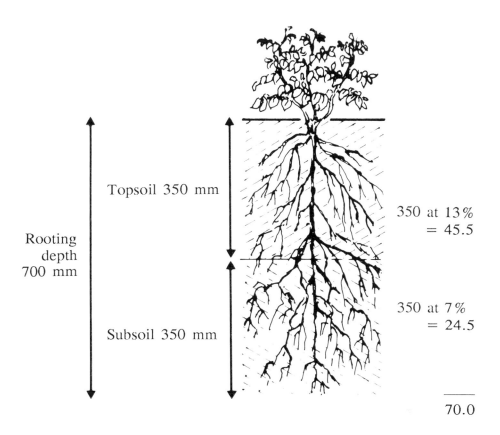

Rooting
depth
700 mm

Topsoil 350 mm

Subsoil 350 mm

350 at 13%
= 45.5

350 at 7%
= 24.5

_____

70.0

Example 2 (see diagram on page 20)

This second example illustrates the more complex calculation required for crops that show a concentration of roots in the upper horizons, and a sparser rooting system below. Consider a crop of winter wheat grown on a similar soil. From Table 2.6, winter wheat can extract all available water to a depth of 800 mm, and easily available water to a total depth of 1,200 mm. From Table 2.3, the 350 mm of topsoil will release 13 per cent AWC, representing an amount of 45.5 mm. As the crop can extract all available water to 800 mm, there will be 450 mm of subsoil that can be fully exploited, and a further 400 mm that will release easily available water only. The 450 mm has 7 per cent AWC, which is 31.5 mm water. The final 400 mm will only

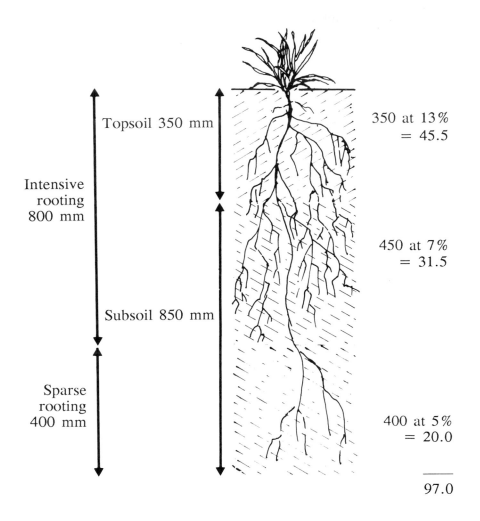

Intensive rooting 800 mm

Topsoil 350 mm

350 at 13%
= 45.5

Subsoil 850 mm

450 at 7%
= 31.5

Sparse rooting 400 mm

400 at 5%
= 20.0

‾‾‾‾
97.0

release 5 per cent water (easily available) which amounts to 20 mm. The total amount of water available to the winter wheat crop is thus 45.5 + 31.5 + 20 = 97 mm.

Example 3 (see diagram on page 21)

Finally, consider cauliflowers grown on a slightly heavier soil, i.e. 400 mm of medium sandy loam over medium sand. In this case, we have various rooting patterns throughout the season and must repeat the calculation for each stage.

(*a*) Five weeks after emergence:

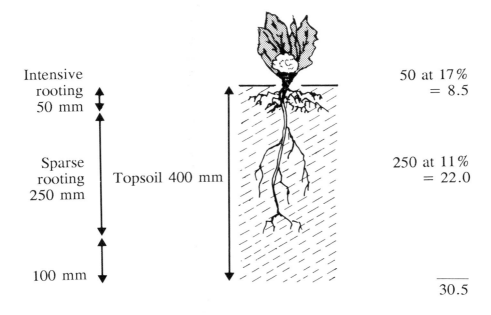

Intensive rooting 50 mm

Sparse rooting 250 mm

Topsoil 400 mm

100 mm

50 at 17% = 8.5

250 at 11% = 22.0

———
30.5

(*b*) Nine weeks after emergence:

Intensive rooting depth: 150 mm
Sparse rooting depth: 700 mm

Horizon (mm)	×	% AWC	=	mm water available
0–150		17		25.5
150–400		11		27.5
400–700		5		15.0

Root zone capacity of cauliflowers at 9 weeks 68.0

(*c*) Eleven weeks after emergence:

Intensive rooting depth: 300 mm
Sparse rooting depth: 700 mm

Horizon (mm)	×	% AWC	=	mm water available
0–300		17		51.0
300–400		11		11.0
400–700		5		15.0

Root zone capacity of cauliflowers at 11 weeks 77.0

(*d*) Fourteen weeks after emergence:

Intensive rooting depth: 500 mm
Sparse rooting depth: 800 mm

Horizon (mm)	×	% AWC	=	mm water available
0–400		17		68.0
400–500		7		7.0
500–800		5		15.0

Root zone capacity of cauliflowers at 14 weeks   90.0

Therefore, in this soil type, there are 30.5, 68, 77 and 90 mm of water available to cauliflowers at 5, 9, 11 and 14 weeks, respectively.

By using this technique, it is possible to calculate the root zone capacity for any crop listed in Table 2.6 and grown on any of the soils listed in Table 2.3.

RELATIONSHIP BETWEEN ROOT ZONE CAPACITY AND CRITICAL SMD

The next step in irrigation planning is to calculate a critical deficit for the crops in question. The precise relationship between critical deficit and root zone capacity is very complex, and varies with crop, soil and prevailing weather conditions, as explained by Doorenbos and Kassam. Some crops, including field vegetables where the harvested part is of low dry matter (e.g. onions, potatoes) are described as needing relatively wetter soils to maintain maximum growth; crops harvested with a high dry matter (e.g. grain crops) can tolerate a greater depletion of the root zone capacity. This is in accord with work in Britain at the National Vegetable Research Station (NVRS) at Wellesbourne (now IHR, the Institute of Horticultural Research), where it was shown that maximum yield of several field vegetables was only obtained if the root zone capacity was never allowed to dry beyond 25 per cent. However, it was also recognised that irrigation at 50 per cent depletion only involved a slight loss of yield, and was more economical and certainly more practical. Aslyng has argued that, under northwest European conditions, including the UK, it can be assumed that production is significantly limited by lack of water only when more than 50 per cent of the root zone capacity has been exhausted. Experience with a range of crops at Gleadthorpe EHF supports this view, and I have adopted it here. For most irrigation planning purposes, therefore, a sufficiently accurate critical deficit can be obtained by using half the root zone capacity. Thus, in the first example already quoted, the critical deficit of the maincrop potatoes is calculated as half of 70 mm, i.e. 35 mm. The soil description used in the example is identical to that at Gleadthorpe EHF, and the estimate of 35 mm is in accord with the critical deficit measured at that site (see Tables 2.1 and 2.2). Similarly, in the example for wheat on this soil, the estimated critical deficit is $\frac{1}{2} \times 97 \simeq 50$ mm, which is again in agreement with field experiments.

This estimation is adequate in hot, sunny conditions but, as already described, in cool cloudy conditions the critical deficit increases to around 70 per cent of the root zone capacity. This means that, in wetter summers, the use of critical deficits based on sunny conditions leads to more irrigation being applied than is strictly necessary. It is difficult to see how this can be overcome because transpiration demand cannot be forecast. It would be possible to adjust the critical deficits, and hence the triggers for irrigation, according to the weather conditions expected during the next week or so, but this could lead to greater problems. If overcast weather is expected, adjusting the triggers accordingly will have the effect of delaying irrigation, with some small economic savings, and the SMDs will increase towards the new higher critical deficits before irrigation is triggered. But if a hot and sunny period then arrived, the critical deficits would have to be adjusted downwards again, and many farmers would find themselves faced with a large number of fields requiring immediate irrigation. This would inevitably result in some fields undergoing drought stress for several days before irrigation could be applied, and a loss of yield or quality might result. If this sequence of events occurred several times in a season, the economic losses could be severe.

The safe approach is to use a critical deficit based on 50 per cent depletion of the root zone capacity. This insures against any losses in a hot and sunny year, but it must be clearly understood that much of the irrigation is wasteful in dull summers.

It is possible to change our strategy and calculate critical deficits based on 70 per cent of root zone capacity, accepting some yield losses in hot and sunny years, but knowing that we are not wasting water in the cool and cloudy years. But the variable costs of irrigating are usually low (£20 per 25 mm per ha at Gleadthorpe in 1988), so the financial losses incurred as a result of applying 50 or 75 mm of unnecessary water in a cool year are greatly outweighed by the benefits of irrigating a high-value crop correctly in a hot and sunny year. The major costs of irrigation are involved with the purchase of capital.

There are a number of important exceptions, however, which should also be considered.

Organic soil

The AWC of organic soils varies between 23 and 45 per cent, depending on the actual soil texture (see Table 2.7).

Table 2.7   Estimation of available water from organic soils

Texture class	Available water %	Easily available water %
Humose sands	23	16
Humose loams	28	20
Humose clays	23	16
Peaty sands	39	36
Peaty loams	27	18
Sandy peats	45	30
Loamy peats	35	26
Humified peats	33	26
Fibrous and semi-fibrous peats	44	35

Source: Soil Survey and Land Research Centre.

Such high AWCs mean that crops are only likely to come under moisture stress at very high soil moisture deficits. However, if these soils are allowed to dry out to their true critical deficits (as calculated here), there are likely to be problems encountered with rewetting of the soil when irrigation is eventually applied. Growers on such soils have learnt from experience not to allow the SMD to exceed 40–45 mm before irrigating, if they are to avoid such problems.

Responsive and non-responsive growth stages

The method of calculating critical deficits as described earlier is designed to produce the maximum growth of the crop, which does not necessarily result in the maximum yield of the harvested portion. For example, oilseed rape, peas and some other crops pass through stages when irrigation will increase the total growth of the crop, but it is unlikely to increase the yield of the fraction that we require, i.e. the seed or fruit. In fact, this desirable fraction may even be decreased by irrigation at the wrong time.

The deficit that is critical for total growth and dry matter accumulation within the whole crop does not always coincide with the deficit that is critical for the production of that portion of the crop that we are seeking to increase. With such crops it is not desirable to keep the soil water depletion below 50 per cent at all times; there are periods during which we can allow it to increase without affecting final yield.

Further details are given in chapters 7 and 8 where the irrigation of peas, beans and oilseed rape, is described.

Low-value crops

As already described, potatoes and field vegetables give high-value returns for irrigation, so a high level of insurance is worthwhile, and the critical deficits for irrigation should be based on a root zone depletion of 50 per cent. The value of cereals, however, is considerably lower compared with the cost of irrigation. A critical deficit based on 50 per cent root zone depletion will probably result in maximum yield in a dry year, but such intensive irrigation could be uneconomic over a run of years, especially if dull summers are experienced. There may be a case, therefore, to adopt a higher trigger deficit for low-value crops such as cereals, and trigger irrigation only after the root zone capacity has depleted by 70 per cent or perhaps more, depending on the value of grain. This will not result in maximum yield in a dry year but could be the most economic option over a long period. A fuller discussion of this can also be found in chapter 7.

Of course, even with a low-value crop such as cereals, if the weather forecasters were to predict with certainty a hot and sunny period, as in 1976 and 1989, then it would be prudent to reduce temporarily the trigger deficit and irrigate at the 50 per cent depletion stage, provided that the water and equipment were available. Then, as soon as the hot period was over, one could return to the more conservative regime based on a higher percentage depletion.

High-quality salad vegetables and fruits

Some crops (e.g. iceberg lettuce) are judged for quality according to the degree of turgidity shown by the produce. This can often be increased by applying water during the weeks leading up to picking. The criteria for irrigation at this time should not be based on 50 per cent depletion of the root zone capacity, but I would advocate irrigating once a depletion of 25 per cent had occurred.

Disease control

In some instances, irrigation can be applied to control a particular disease, e.g. common scab control on potatoes. The 'critical' deficit for such disease control often bears no relationship to the normal critical deficit for drought stress prevention. Details for such irrigation strategies can be found in relevant chapters later in this book.

Future refinements

It has been suggested by some authors that rooting patterns vary between soil types, and crops root more deeply in soils of higher organic matter or clay content. Other authors have argued the converse, believing that crops root to a greater depth in sands. In the absence of a consensus, any such effect has been ignored. When more information becomes available, any variation in root growth could easily be incorporated into the model.

Another possible future refinement involves the degree of lateral movement of water in soils. When an isolated root removes water from the immediately surrounding soil, there is a tendency for water elsewhere to move into this drying area and replace the extracted water. The ability of soil water to move in this way is termed the hydraulic conductivity, and is dependent on both the soil type and the amount of soil moisture present. Where roots are rather sparse in lower horizons, the present assumption is that the crop can only extract the easily available water. This is a rather simplistic view because the amount of water available to such a root system will depend on the hydraulic conductivity of the soil. If the hydraulic conductivity is high, there is a greater potential for water to move towards those soil pockets containing roots, thereby replacing the extracted water, and enabling a much better exploitation of the available water reserves throughout the soil profile.

The need for such refinements must, however, be placed in perspective. It is important to understand how irrigation timing is affected by the calculation of critical deficit. The critical deficit determines the timing of the first irrigation. As an example, consider a situation in which the SMD builds up by 25 mm per week and each irrigation applies 25 mm. If 25 mm were used as a critical deficit, irrigation would start at the end of the first week, and be repeated at weekly intervals. If, however, 50 mm was used as a critical deficit, the first irrigation would be delayed until the end of the second week, but irrigation after that would be repeated at weekly intervals. The choice of critical deficit for irrigation affects the date on which water is first applied, but has little effect after that.

Thus any refinement that is likely to change the critical deficit by 5 mm or less is unlikely to have a greater effect than delaying or bringing forward the start of irrigation by one or two days. It is important to estimate critical deficits with reasonable accuracy, however, as experiments have shown that the correct timing of this first application is important for both yield and quality.

Growers and advisers should find that the technique suggested in this chapter will produce estimates of critical deficit which are precise enough for most practical situations.

SUMMARY

The calculation of critical soil moisture deficit is a complex subject and it is useful to summarise it again here.

It is first necessary to describe the soil texture of both topsoil and subsoil. For this, soil examination is essential. This can be done by particle size analysis, but for an experienced adviser it is usually adequate to use hand-texturing, i.e. from the feel of moist soil between finger and thumb. Having established the soil texture, the AWC of topsoil and subsoil can be determined by reference to Tables 2.3 and 2.4. Where stones are commonly found, there is a need to adjust the AWC accordingly.

This is fairly straightforward, i.e. 20 per cent stones by volume will require a reduction of 20 per cent in the AWC, etc.

As a broad rule of thumb, the critical deficit can be assumed to be 50 per cent of the root zone capacity, but this will vary in some situations. When calculating root zone capacity, it is fair to assume that the abstraction of water approaches 15 bar suction where the highest concentration of roots is found, but only that available to 2 bar suction should be used where root growth is limited. Rooting patterns for use in this context are shown in Table 2.6.

The importance of critical SMD in irrigation planning has already been emphasised. Using the methods described in this chapter an adviser or grower should be able to calculate a critical SMD for most of the situations that he is likely to encounter.

REFERENCES AND FURTHER READING

ASLYNG, H. G. (1984), 'Soil water capacity, climate and plant production. The development and future of irrigation', *Proceedings of the North-Western European Irrigation Conference*, Billund, Denmark, 2–39.

BARRACLOUGH, P. B. and LEIGH, R. A. (1984), 'The growth and activity of winter wheat roots in the field: the effect of sowing date and soil type on root growth of high yielding crops', *Journal of Agricultural Science, Cambridge 103*, 59–74.

BREWSTER, J. L. (1977), 'The physiology of the onion', *Horticultural Abstracts 47*, 17–23.

DOORENBOS, J. and KASSAM, A. H. (1979), 'Yield response to water', *FAO Irrigation and Drainage Paper No. 33*, pp. 193.

DOORENBOS, J. and PRUITT, W. D. (1977). 'Crop water requirements', *FAO Irrigation and Drainage Paper No. 24*, pp. 144.

DUNHAM, R. J. (1987), Irrigating sugar beet in the United Kingdom', *Proceedings of the 2nd North-Western European Irrigation Conference*, Silsoe, UK.

GARWOOD, E. A. (1987), 'Water deficiency and excess in grassland: the implications for grass production and for the efficiency of use of N', *Proceedings of a conference on 'Nitrogen and Water Use by Grassland'*. North Wyke Research Station, Devon, UK, 24—41.

GARWOOD, E. A. and SINCLAIR, J. (1979), 'Use of water by six grass species. 2. Root distribution and use of soil water', *Journal of Agricultural Science, Cambridge* 93, 25—35.

GREENWOOD, D. J., GERWITZ, A., STONE, D. A. and BARNES, A. (1982), 'Root development of vegetable crops', *Plant and Soil 68*, 75—96.

HALL, D. G. M., REEVE, M. J., THOMASSON, A. J. and WRIGHT, V. F. (1977), 'Water retention, porosity and density of field soils', *Soil Survey Technical Monograph No. 9*.

HEATH, M. C. and HEBBLETHWAITE, P. D. (1985), 'Agronomic problems associated with the pea crop', in P. D. Hebblethwaite, M. C. Heath and T. C. K. Hawkins (eds), *The Pea Crop — A Basis For Improvement* (Butterworth), 19—29.

HEBBLETHWAITE, P. (1982) 'The effects of water stress on the growth, development and yield of *Vicia faba* L', in C. Hawtin and C. Webb (eds), *Faba Bean Improvement*, 165—75.

HIGGS, K. H. and JONES, H. G. (1989), 'Water use by strawberry in south-east England', *Journal of Horticultural Science 64* 167—75.

HOLLIS, J. M. (1987), 'The calculation of crop-adjusted soil available water capacity (AP) for wheat and potatoes', *Soil Survey Research Report No. 87/1*.

HUGHES, H. M. (1965), 'Strawberry irrigation experiments on a brickearth soil', *Journal of Horticultural Science 40*, 285—96.

STEGMAN, E. C. (1983), 'Irrigation scheduling: applied timing criteria', in D. Hillel (ed.), *Advances In Irrigation Volume 2* (Academic Press, New York), 1—30.

Chapter 3

Estimation of Irrigation Requirement and Expected Yield Response

When considering irrigation investment on a farm, it is essential to have an estimate of the water requirement and the likely responses that can be expected from irrigation. To obtain such estimates, growers and advisers have commonly used the results of a group of irrigation experiments from a research station working with the relevant crop. This is a wrong approach to adopt and will, more often than not, lead to false conclusions. For instance, the research station and the farm in question may have different soil types, or they may be regularly subjected to different weather conditions.

In the UK, for example, the climate is such that western districts receive much more rainfall than those on the eastern side of the country. The irrigation requirement measured at a research station in the east is not much help to growers in the west, unless the climatic variation is taken fully into account. Irrigation requirement is also dependent on the amount of sunshine received by a crop. This also varies with locality, the highest amounts being found in southern districts of the UK. The broad picture resulting from these effects can be seen in Colour plate 3, which shows the SMD attained during summer months in different parts of England and Wales. It must be noted that this is a reflection of the climatic variation, and is independent of soil type.

Even if a farm is situated close to a research station, and conditions are similar in all respects, it would be misleading to take the results of the experiments if they were only conducted over a small number of years. The weather conditions prevailing during a sample of years may not necessarily represent the long-term weather pattern at the site.

As an example, consider the yield response from irrigating Pentland Crown potatoes at Gleadthorpe EHF over the period 1966–84. In the first nine years (1966–74), irrigation produced an average yield increase of 6.2 t/ha. There is no data for 1975 (due to damage resulting from a devastating hailstorm), but during the next nine years the average response shown by the same variety of potatoes was 17.2 t/ha. The yields are shown in Figure 3.1.

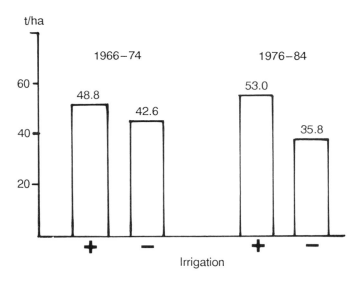

Figure 3.1   Yields of Pentland Crown, 1966–84.

During the years 1966–74, irrigation increased the yield from 42.6 t/ha to 48.8 t/ha. During the years 1976–84, the irrigated yield was higher, at 53.0 t/ha, but the unirrigated yield was considerably lower, at 35.8 t/ha. There was no change in husbandry that could account for the difference, nor is there any reason to suspect that too little irrigation was applied in the first series. Indeed, the large response in recent years is as much due to lower yields of the unirrigated crop as anything else. So why have the returns from irrigation been higher in the recent period? The most likely explanation lies in the variation in Britain's summer weather.

The response of a potato crop to irrigation largely depends on the amount, and distribution in time, of the rain that the crop receives, and daytime temperatures. These parameters for the two nine-year periods 1966–74 and 1976–84 are shown in Table 3.1, along with the potential soil moisture deficit (PSMD).

Potential soil moisture deficit (PSMD) is a fairly good indicator of drought stress. The concept of SMD has already been explained in chapter 2. As a soil dries towards the critical SMD, the actual rate of transpiration falls below the potential rate, and the SMD increases more slowly. If, however, we considered a hypothetical situation of a crop that could continue to transpire water at the potential rate, a very high SMD would develop in dry conditions. This potential soil moisture deficit (PSMD) can be calculated from weather data and,

Table 3.1   Weather records at Gleadthorpe EHF

	1966–74	1976–84
Average accumulated summer temperature (°C)	2,273	2,318
Average summer rainfall (mm)	247	195
Average PSMD (mm)	131	193
Years with PSMD greater than 200 mm	1970	1976
		1977
		1981
		1983
		1984

Source: Gleadthorpe EHF, ADAS.

although it is only a theoretical figure, it represents a good measure of the drought.

In the years 1966–74 the mean summer (six months) accumulated temperature was 2,273 day °C, but in 1976–84 this increased to 2,318 day °C. This difference of 45 day °C may not appear large, representing only one-quarter of a degree per day, but measured over such a long period it is a significant increase. Average summer rainfall is interesting as it was markedly different, with 247 mm for the earlier nine years, but only 195 mm in the more recent period.

In 1966–74, the average value of the PSMD was 131 mm, but in 1976–84 it was considerably higher at 193 mm. A PSMD above 200 mm can be considered to represent a high drought stress on sandy soils. In the period 1966–74 this occurred in only one year, but in the period 1976–84 it occurred in five years. It can be concluded that droughts have occurred more frequently in recent years than in the previous period. Most of us remember the extreme weather conditions of 1976, but it is interesting to note that there have been many other drought years in the recent period, and this is undoubtedly why responses to irrigation were much higher.

The lesson here is straightforward. A study of irrigation responses at Gleadthorpe over the relatively long period of 1966–74 could easily lead to the false conclusion that the irrigation requirement of potatoes and the resulting yield responses are rather low. Similarly, a study of 1976–84 in isolation could lead to the equally false conclusion that the irrigation requirement and responses are very high. Investment in irrigation today depends on the yield responses expected in future years. These are particularly difficult to forecast from a small set of experiments because British summer weather is so variable, and even a lengthy series of experiments, lasting nine or ten years, may not be sufficient to provide an accurate picture. In order to calculate irrigation requirement and estimate an expected yield response on a particular farm, it is necessary to interpret the results from irrigation experiments against the background weather in which they were obtained. This information can then be modified with reference to the *long-term* weather pattern at the farm for a period of at least twenty years.

Fortunately, the Meteorological Office in the UK have a comprehensive system of weather stations that have been collecting data for many years. With access to this data, it is possible to look at long-term weather patterns within any particular locality and, together with the results from the irrigation research at experimental centres, estimate irrigation requirements and expected yield responses.

To do this, a logical sequence of steps must be followed. First at the research stations:

1. Experiments must be conducted to determine which stages of growth are likely to be responsive to irrigation. The results for each crop, where known, are described in the relevant chapters later in this book.
2. Experiments are also conducted to determine the critical SMDs for each crop during the important stages of growth, and these are related to root zone capacity of the soil. These results, where available, are also presented in the chapters on specific crops.
3. The same experiments are used to determine the response obtained for each unit of irrigation applied. These are known as *crop response functions*.

The next set of steps involve taking all these experimental results and applying them to the specific farm situation:

4. The critical SMD is calculated for each crop, based on the known properties of the soil type on the farm, and the known requirements of the crop from experimental results.
5. The amount of water required to irrigate each crop on the farm is calculated, based on the calculated critical SMD, taking into account the weather conditions that are likely in that locality.
6. The crop response functions obtained at the research stations are then applied to the calculated amount of irrigation required on the farm, and a likely yield response is obtained.

The first three steps fall into the province of the research worker, but every farmer or adviser planning irrigation investment should be familiar with the latter three steps.

CALCULATION OF AVERAGE IRRIGATION REQUIREMENT

In order to calculate the average irrigation requirement for a particular soil and site, it is first necessary to examine the soil and calculate a critical soil moisture deficit for each crop, as described in the previous chapter, taking into account the individual crop requirements detailed in the relevant chapters on each crop.

The next task is to study the weather data for the site over a large number of years (at least twenty years) and calculate how much irrigation was necessary for each crop in order to keep SMDs below the critical deficit in each year. A balance-sheet approach can be used, with rainfall as a credit, and evapotranspiration of the crop as a debit, and a record is logged of every occasion on which irrigation would have been required in order to keep the SMD below the critical deficit. This could be a time-consuming and complex task, but in several European countries (e.g. Denmark, West Germany, East Germany and the UK), facilities have been developed that make it relatively easy.

In the UK the Meteorological Office initially published an atlas showing the irrigation amounts required in different areas of the UK to keep below certain SMDs. This publication is now outdated owing to the much greater quantity of meteorological data since available, and the more sophisticated analysis that is now possible with computers.

A computer program to conduct this analysis has now been developed jointly by ADAS and the Meteorological Office. The computer requires details of soil type, critical deficits, critical stages of growth and certain agronomic data for each crop, e.g. average date of planting, average date of 25 per cent and 75 per cent crop cover, etc. The computer has access to weather data over the period 1961–80 from several hundred weather stations, based on a key network of ninety-four weather stations in England and Wales which have complete records. This specific period has been chosen for several reasons. There is a very comprehensive set of meteorological data throughout this period. It is also believed that these years include the whole range of summer weather that could be expected in the near future, providing information on a number of 'typical' years, along with extreme summers such as 1976. Using the nearest station to the farm, the weather from each year is analysed. At the end of the analysis, the amount of irrigation required for each crop in each of the years 1961–80 is produced.

This type of analysis is the only accurate way to estimate the average irrigation requirements of various crops on a *particular* farm because it uses a typical twenty-year weather pattern from a weather station near to the farm. Furthermore, as the critical deficits used are based on examination of the soil on the farm, the results are tailored to fit that *particular* situation.

ESTIMATING CROP RESPONSES

Having established the irrigation requirement for each crop on the farm, it now remains to estimate the likely responses from irrigation. These are obtained by applying the crop response functions to the estimated irrigation requirement.

The crop response functions have been obtained from experiments. In some cases, e.g. potatoes and sugar beet, many experiments have been carried out and much confidence can be placed on the response functions. For many crops, however, there is a shortage of data so it has been necessary to estimate response functions from relatively few trials.

The MAFF/ADAS *Reference Book 138* 'Irrigation' quotes response functions for a range of crops. These figures are presented in Table 3.2, after some adjustments to take into account the results of ADAS research since publication.

It must be emphasised that these figures are the average response to irrigation expected over a long run of years, incorporating the low value of irrigation in some years due to rain falling subsequently (see Figure 3.2). This means that the response to irrigation in a dry year would be higher than the figures quoted in Table 3.2, and conversely the response would be lower in a wet year.

Although the figures are quoted as yield responses per mm of irrigation applied, it must be understood that yields will not rise indefinitely if more and more irrigation is applied. These figures represent the irrigation responses resulting from good irrigation practice, i.e. that in which the irrigation amounts applied are sufficient to keep SMDs below the critical SMD for each crop, but no more.

In the previous section, I described how the ADAS/Meteorological Office computer program can be used to calculate the average irrigation requirement for a specific farm. Provided the correct critical deficits and irrigation regimes are used, the calculated average irrigation requirement will be that which relates to good irrigation practice, and so can be used to calculate an average yield response. It is a simple calculation to multiply the average irrigation requirement by the yield responses in Table 3.2 and estimate the average long-term yield response for that particular site.

In order to provide some examples, the program has been run for

Table 3.2   **Average yield responses (t/ha) from each mm of irrigation applied**

Crop			Yield response (t/ha mm)	
Agricultural crops				
Cereals			0.007	
Grassland			0.025	(dry matter)
Peas	—	vining and dried	0.04	(first 25 mm)
			0.01	(subsequent applications)
Potatoes	—	early	0.05	
Potatoes	—	second early	0.06	
Potatoes	—	maincrop	0.09	
Sugar beet			0.09	
Vegetables				
Beans broad	—	processing	0.04	
Beans French	—	freezing	0.06	
Beans runner			0.05	
Brussels sprouts	—	early	0.04	
Cabbage	—	summer	0.14	
Carrots	—	early	0.07	
Carrots	—	late	0.13	
Fruit				
Apples	—	Cox	0.015	
Blackcurrants			0.03	
Raspberries			0.025	
Strawberries			0.025	

Source: MAFF, ADAS.

Gleadthorpe, Terrington and Arthur Rickwood EHFs, and the results are shown in Table 3.3.

The ADAS/Meteorological Office program also provides one further set of data that has not yet been described. It calculates the amount of water required in each year between 1961 and 1980, and also estimates the date of each successive application. From this, one can determine whether a 5-day cycle, 7-day cycle or 10-day cycle is necessary to achieve the irrigation requirements of each year. This is essential knowledge to determine the amount of application equipment required.

As an example, consider a situation where a 10-day cycle has been shown to be sufficient for the maincrop potatoes on a farm. If there are 30 ha of potatoes on the farm, equipment that could irrigate 3 ha per day would be required. If, however, the required cycle time was 5 days, it would be necessary to irrigate 6 ha per day, and the grower would have to provide equipment with greater capacity. Armed with this data, the required number and size of hosereels, etc. can

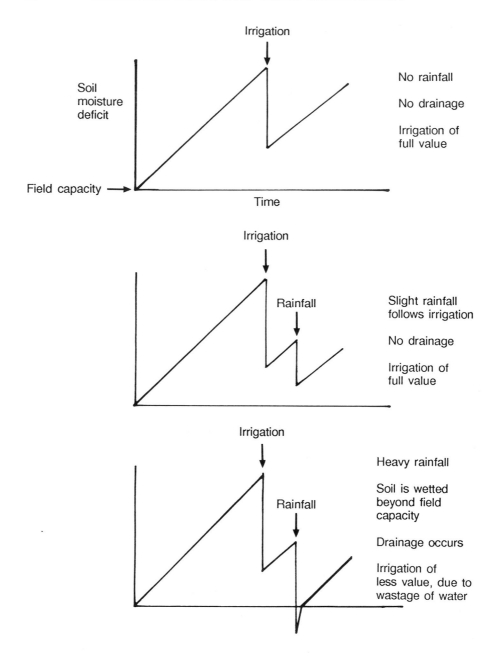

Figure 3.2   Diagrammatic representation of the decreased value of irrigation in the event of subsequent heavy rainfall.

Table 3.3 Estimates of average irrigation requirement and yield response

Farm	Crop	Average irrigation requirement (from ADAS/Met. Office computer program) (mm)		Yield response per mm (from Table 3.2)		Long-term average yield response (t/ha)
Gleadthorpe	Potatoes	111	×	0.09	=	10.0
	Sugar beet	49		0.09		4.4
	Cereals	25		0.007		0.2
Terrington	Potatoes	45		0.09		4.1
	Sugar beet	4		0.09		0.4
	Cereals	0		0.007		0.0
Arthur Rickwood	Potatoes	133		0.09		12.0
	Sugar beet	25		0.09		2.3
	Cereals	1		0.007		0.0

Source: ADAS.

be calculated with confidence, greatly assisting accurate investment appraisal.

ESTIMATION OF THE RANGE OF RESPONSE

So far in this chapter I have described how *average* irrigation require-ments and yield responses can be estimated. It has been common practice to assume average responses each year when conducting a financial appraisal, but this ignores the important influence that year-to-year variation can have on the value of irrigation. It is not my inten-tion here to describe financial business management techniques, but it is useful to discuss how an estimate of this variation can be obtained.

The ADAS/Meteorological Office program, as mentioned previously, provides not only an average irrigation requirement, but also an indi-vidual requirement for each year between 1961 and 1980. Using the crop response functions in Table 3.2, the irrigation response could also be estimated for each of those individual years. This will give an approx-imate idea of the range of expected irrigation requirements and responses over a long period.

The estimates of response for individual years are not very precise because they do not take account of the principle illustrated in Figure 3.2, namely, that each individual application of water may have far less effect in some years due to heavy rain falling subsequently.

It is possible that the crop response functions could be further refined by using different functions each year, depending on the weather. For instance, instead of using 0.09 for potatoes in every year, a slightly better distribution of responses can be obtained by using 0.01 in wet years and 0.11 in medium and dry years. This has the effect of extending the range of responses during the period, and giving a slightly better forecast of future responses. However, there is a scarcity of data on the functions to use for most crops, and work to produce more refined response functions is under way. In the meantime, however, using the simple crop response functions presented here does give an appreciation of the range of responses that are likely on each particular site.

An example is given in Table 3.4, which uses the irrigation requirement for maincrop potatoes at Gleadthorpe EHF as an example. Multiplying the requirement in each year by 0.09 (from Table 3.2) provides an estimate of yield response in each year.

The average yield response is estimated to be 10.0 t/ha which agrees well with that actually measured over the period (9.3 t/ha), as well as longer term measurements (10.6 t/ha). Using the simple function in both wet and dry years has narrowed the response range, producing small estimates in years of nil response, but also underestimating the response in 1976. The data, however, provide a better representation of the likely returns from irrigation than can be obtained by considering only the average response. With the meteorological dataset now available, a table of this type can be produced for any particular farm.

The final step of the analysis involves setting the cost of irrigation against the returns over the period. In some years, such as 1976, the irrigation requirement will be high, and a great deal of financial investment in equipment would be required to supply the required amount of water to the crop. Providing for such an extreme is unjustified, and it is better to aim at providing the requirement corresponding to the fourth or fifth driest year in the period. The estimated responses in years such as 1976 would then have to be modified downwards somewhat, to make allowance for the restricted irrigation available.

It is then a matter of calculating the cost of investment, and balancing it against the financial return. Much of the financial benefit arises from the yield increase, but likely higher prices obtained in the drier years should also be taken into account. In addition, when assessing the financial returns, it is important to take account of any improved quality that could result from irrigation, as this can even outweigh the benefits resulting from the yield increase. Consider common scab control of potatoes (see chapter 5). This disease can markedly suppress the price obtained for the crop. Irrigated and unirrigated potatoes at Gleadthorpe in 1988 differed in value by approximately £20 per tonne, largely as a result of this disease. The financial benefit from irrigation was far greater than the 2.9 t/ha extra yield that was obtained in 1988!

Table 3.4 The estimation of maincrop potato yield response from irrigation requirement data at Gleadthorpe EHF

Year	Irrigation requirement for maincrop potatoes (mm)	× 0.09 =	Estimated yield response (t/ha)
1961	100		9.0
1962	150		13.5
1963	100		9.0
1964	125		11.3
1965	25		2.3
1966	100		9.0
1967	150		13.5
1968	50		4.5
1969	100		9.0
1970	200		18.0
1971	50		4.5
1972	75		6.8
1973	25		2.3
1974	100		9.0
1975	200		18.0
1976	275		24.8
1977	175		15.8
1978	75		6.8
1979	125		11.3
1980	25		2.3
Average			10.0

Source: ADAS.

A further refinement is possible. The twenty years' pattern of data can, if required, be rearranged in various combinations to examine the effect on cashflow of, for example, the driest years occurring at the start of the period, or conversely, not until the end, although given the impossibility of forecasting it is doubtful if there is any benefit in doing so. It would be highly dangerous to invest on the basis of a preponderance of dry years early on.

Some situations are clear cut with regard to irrigation investment. For instance, high-value horticultural crops grown on sandy soils can usually justify the investment easily. Conversely, most crops grown in areas of high rainfall are unlikely to justify the investment. The analysis described here should help with the large majority of situations that fall between these extremes.

There is, however, a further important factor to discuss. Many growers invest in irrigation even though a conventional farm business management exercise shows it to be unprofitable in the long term. Why? Because they look upon it as an essential insurance against the disastrous effects of a severe drought in an occasional year. This applies particularly to high-value field vegetables, and also to potatoes grown for the high premium quality market.

In this respect, irrigation is like any other insurance policy. We spend money on insurance, not because a reasoned financial appraisal shows that a net profit is likely, but because there is a need to protect ourselves against catastrophic events, even though the probability of such events may be quite small. Likewise, many growers invest in irrigation, in the knowledge that it is unlikely to show a net profit in the long-term analysis, but because the enterprise cannot bear the effects of a severe drought.

This is the explanation for the upsurge in irrigation investment immediately following the drought of 1976. Most growers appreciated that such a drought was unlikely to be repeated for a long time, but they also realised their vulnerability in the unlikely event of it recurring. A thorough investment appraisal must take into account the likely effects of such years on the farm business.

The procedures described in this chapter are complex, and usually require the assistance of agronomists and farm business management advisers. However, they are essential if an investment appraisal is to be based on a realistic assessment of the likely economic costs and returns from irrigation on any particular farm. All too often in the past such investments have been based on little more than guesswork.

REFERENCES AND FURTHER READING

BAILEY, R. J. and MINHINICK, J. (1989), 'The agricultural requirements for irrigation water', *Journal of the Institution of Water and Environmental Management 3 (5)*, 451–8.
MAFF (1982), 'Irrigation', *Reference Book 138*.

Chapter 4

Irrigation Scheduling

This book is largely concerned with the agronomic aspects of irrigation, and it is not intended to deal with application equipment. However, there is a need to briefly summarise the types of equipment available, because it has some influence on our choice of irrigation scheduling method.

DISTRIBUTION EQUIPMENT

Spraylines

A sprayline is a length of piping (up to 200 metres) with nozzles inserted, usually at intervals of about 75 cm. Water is sprayed up to a distance of 7—8 metres either side of the line. There are two basic types:

1. Oscillating spraylines, which are driven by water pressure on water motors.
2. Static spraylines, which have multiple jets at different angles.

Moving spraylines about is labour intensive and their use is now largely restricted to salad crops, flower crops and small-scale production of vegetables.

Rotating sprinklers

This type of sprinkler consists of a spring-loaded head, containing one or two nozzles, which rotates and produces a circular wetted area. The sprinklers are attached along lateral pipes so that the circular wetted areas overlap. There are three basic systems:

1. Conventional sprinklers. The whole assembly of lateral pipes and sprinklers is moved every few hours to the next area to be irrigated.
2. Reduced sprinkler system. The sprinklers are attached to the laterals via a valve connection which automatically opens when the

sprinkler is inserted, and closes again when the sprinkler is removed. Enough laterals are laid out for a complete day's irrigation, so that only the sprinklers have to be moved every few hours. This is a considerable saving on labour throughout the day.
3. Solid-set sprinkler systems. The reduced sprinkler system is taken one stage further. Enough laterals to irrigate the whole crop are placed in position for the duration of the season, and the sprinklers are moved as required. This consumes less labour than any of the systems previously described, but demands far more capital.

Mobile sprinkler lines

These include systems where conventional sprinklers are used, but the movement of the lateral pipes is mechanised. The lateral may be fixed to a set of wheels which are static when irrigation is in progress, but can be driven to the next location when irrigation is complete. Alternatively, sprinklers can be fitted to a large lay-flat hose which can be laid out and retrieved using a tractor.

Travelling rainguns

These irrigate continuously as they traverse the field, producing a wetted strip between 40 and 100 metres in width (see Colour plate 4). With the early designs, the raingun was self-propelled, winching itself along a wire using a waterdriven mechanism. Later designs (available in the UK since 1974) consist of the raingun being pulled in by its supply hose which is slowly wound onto a large hosereel (see Colour plate 5).

All these machines are subject to poor distribution patterns when used in windy conditions. They are, however, popular due to their low labour requirement compared with most other systems.

Hosereel booms

These operate in a similar way to the hosereel machines described previously but, instead of the water issuing from a raingun, it is distributed via a travelling boom (see Colour plate 6). There are two basic types:

1. Booms with sprinklers spraying upwards. These are subject to uneven distribution in a wind, but considerably less so than rainguns.
2. Booms with nozzles spraying downwards. These produce a very uniform distribution pattern even when used in relatively windy

conditions. Unfortunately, they have a high instantaneous application rate and are liable to cause run-off. They are better suited to applications of 12 mm rather than 25 mm. This is probably ideal for seedbed irrigation, many vegetable crops in the early stages, and also for common scab control on potatoes (see chapter 5).

Centre-pivot and linear irrigators

These machines have become extremely popular in some areas of the world, because of their minimal labour requirement and the very even distribution patterns obtained with them. Basically they consist of a large moving gantry which supports sprinklers or drop nozzles. Centre-pivoted machines rotate in a large circle covering one or more fields. Linear machines are similar in concept, but move in a straight line, irrigating a large rectangular area (see Colour plate 7).

It is possible to combine the two designs, with a machine that will move in a straight line, irrigating a wide strip, but then pivot through 180 degrees at the end of the run and return along an adjacent parallel strip.

All of these machines can be left automatically to irrigate very large areas.

Trickle systems, micro-sprinklers and mini-sprinklers

This category consists of various types of equipment, all expensive, but which can be more efficient in wide row crops, as found in orchards. They involve a minimal labour requirement, and use less water than the previously described systems because the water is only applied where required, not across the total area of the field.

Trickle (sometimes called drip) irrigation can take the form of surface or underground piping laid in rows, with spaced emitters or continuous seepage of water from the pipe.

On sandy soils with limited potential for lateral flow of water, the soil may become wetted only in very narrow strips. This can cause problems because the water may reach only a limited proportion of the roots. In 1976, a very dry year, many orchards using trickle irrigation suffered drought stress because of this. To avoid this, it is necessary to consider carefully the required number of laterals and outlets relative to soil type.

It can be seen from the above discussion that the choice of irrigation distribution equipment is a wide one, and involves careful consideration of:

1. Cost.
2. Availability of labour.
3. Required accuracy and uniformity of irrigation.

Each of the systems described has its place, because every farm is a unique situation with a different combination of crops grown, area to be irrigated and available labour. Similarly, as can be seen from the following discussion, different situations require different approaches to irrigation scheduling.

IRRIGATION SCHEDULING

The increased yields and improved quality obtained by irrigating crops at Gleadthorpe EHF and other research centres demonstrate the benefits to be derived from irrigation on light soils. To obtain these benefits, however, considerable effort has to be spent to establish the correct amounts and timings of irrigation. The water is not applied on the day that the soil simply appears to be dry, nor on the day that happens to be most convenient, but the correct amount is applied at the correct time, based on an understanding of each individual crop's requirements and the practicalities of application. This careful planning is usually referred to as 'irrigation scheduling'.

Farmers and growers sometimes comment that the irrigation responses obtained at Gleadthorpe are higher than they obtain themselves, even though their soil type and climate are similar. This may be due in part to their having restricted facilities for applying water, but very often it is the result of poor irrigation scheduling.

Some growers make little attempt at scheduling, but apply irrigation as soon as possible after they have seen one of their neighbours irrigating! This is unsatisfactory because it depends entirely on the assumption that the neighbour is scheduling correctly.

Some growers rely on a cursory soil examination to determine when they should irrigate. This may be better than relying on a neighbour, but is still rather crude. It is possible to examine the soil and decide that irrigation is necessary but, commonly, the crop will suffer from moisture stress before the soil reaches such an obviously dry state. The aim of irrigation scheduling is to apply the water *before* the soil becomes dry enough to affect the crop, and this is impossible to do by visual examination.

MEASUREMENT OF SOIL MOISTURE

Soil moisture content can be determined accurately by sampling the soil, weighing it, drying it in an oven and weighing it again. This is such a time-consuming technique, however, that it is rarely used even in irrigation research, where very accurate estimates of soil moisture are required. It is probably never used to schedule irrigation on a farm scale.

In the early 1950s, scientists in Canada and the United States described the possible use of neutron radiation in measuring the water content of soil. Since then, portable instruments known as *neutron probes* have been developed for this technique. The technique is based on the fact that fast-moving neutrons emitted by a radioactive source are slowed down when they come into contact with water. With a detector capable of distinguishing and counting these resulting slow neutrons, it is possible to estimate how much water is in the soil.

The usual method is to drill a hole, insert a lining of suitable material and lower a probe into the hole. The probe consists of a radioactive source of fast neutrons and a detector of slow neutrons. Readings are taken at various depths and a soil moisture profile is produced.

The technique is widely considered to be very accurate, but has drawbacks. The major one is that it is a point measurement that relates only to the immediate vicinity of the access hole. Where water application is uniform, as is the case with much research work, it is fair to assume that the point reading represents a much larger area. If water is applied unevenly, as is usually the case when using rainguns, especially in windy conditions, an accurate measurement at one point tells us very little about the state of the rest of the field. This can be overcome by having a number of access holes but, because of the variability of water application achieved in some situations, many points would be required to give a good overall picture of the moisture status of the field. In order to obtain the best value from the use of a neutron probe (or any other point measurement), it is essential that irrigation be applied evenly.

A second drawback is that the slow neutron count obtained requires careful calibration to transform it into a moisture content and, if many readings are involved, it requires the use of a computer. This is no problem to the research worker and, with the increasing use of microcomputers nowadays, may not pose a problem on many commercial farms. However, interpretation of the moisture profile, and using it as a management tool to predict when irrigation will be necessary, is a specialist job.

Lastly, but certainly not least, it is doubtful whether a host of farmers could obtain a licence to keep, or would even wish to keep, a radio-active source on their premises. This last factor, together with the requirement for specialist skills to interpret the measurements, means that the technique is best used by a contractor. One such commercial service has been operating in England since 1984. The scheduling advice is based on a computerised water balance technique (see below), with regular checking and updating by neutron probe data obtained during visits by specialist staff.

TENSIOMETERS

Tensiometers are instruments that measure soil moisture tension, i.e. the tension with which water is held by the soil. This is a particularly useful measurement, as it is directly related to the ability of the crop to extract water from the soil at the time of the reading, and is not dependent on the knowledge of previous rainfall, irrigation or weather conditions. Of all soil moisture measurements or calculations, this is the one that is most closely related to the soil moisture situation experienced by the crop.

It again has the advantage of being a direct measurement of the farmer's own crop, and in his own soil.

A tensiometer is basically a porous cup containing water, which is placed in the soil. As the soil becomes dry water moves out of the cup creating a vacuum elsewhere in the instrument, and this can be read from a dial gauge (see Colour plate 8).

The instruments are normally sited in pairs, one being placed at about one-third of final rooting depth, and the other within the lower third of the rooting zone. The principle behind their use is that each crop has a critical tension independent of soil type, and irrigation should be applied before this tension is recorded in the upper instrument.

The lower instrument helps to determine how much irrigation is required to restore the whole rooting profile to a moist state, but avoiding the risk of over-watering.

It must be pointed out, however, that tensiometers suffer the same drawback as other point measurements. When using rainguns in windy conditions, the uneven application rates achieved may lead to local errors which could, in turn, give a erroneous picture of the soil moisture status throughout the rest of the field. However, in other situations, such as when using trickle irrigation, polythene tunnels, or when a fluctuating water table occasionally rises to within the rooting zone, tensiometers have advantages over other techniques.

It is important to pay careful attention to the siting of the instruments, which should be in a representative part of the field. This will ensure that the instruments reach the critical tension, and thus trigger irrigation, when the bulk of the field requires irrigating. If this siting is done badly, and the instruments placed in a particularly moisture-retentive soil pocket, or in a hollow that floods, irrigation will not be triggered until after the bulk of the field has experienced moisture stress.

CALCULATION OF SOIL MOISTURE USING THE PENMAN EQUATION

The major advance in irrigation scheduling occurred in 1948, when H. L. Penman published an account of how to calculate the potential transpiration of a crop from a set of meteorological variables. It was claimed that this calculation was reasonably accurate on a field scale and, further, was independent of the type of crop, provided that it was short, green and covered the ground completely.

The method was based on the principle that the incoming solar radiation reaches the surface of the earth and is used in various ways, which include warming the air, warming the soil, plant growth processes, and providing the energy for evaporation and transpiration. The amounts required for warming the soil (2 per cent) and plant growth (1 per cent) are small enough to be ignored, so the energy absorbed by the earth can effectively be split between that used to heat the air, and that used for evaporation and transpiration. The total amount of energy absorbed can be calculated from easily measurable meteorological data, and aerodynamic theory is used to separate this between that used to warm the air and that used for evapotranspiration.

In recent years, the method has been refined by Monteith and others and is now used to calculate evaporation and transpiration accurately. The equation is, of course, still too complex to be used on a regular basis by farmers and growers. It consists of:

$$\lambda E = \frac{\Delta (R_N - G) + PC_p (e_s - e) / r_a}{\Delta + \gamma (1 + r_s / r_a)}$$

where E = rate of water loss (kg m^{-2} s^{-1})
  Δ = rate of change of saturated vapour pressure with temperature (mb°C^{-1})
  R_N = net radiation (Wm^{-2})
  G = soil heat flux (Wm^{-2})
  P = air density (kg m^{-3})

C_P = specific heat of air at constant pressure (1005 J kg^{-1})
e_s = saturation vapour pressure at screen temperature (mb)
e = screen vapour pressure (mb)
λ = latent heat of vaporisation ($\simeq$2465000 J kg^{-1})
γ = psychromatic constant (= 0.66 for temperatures in deg. C and vapour pressures in mb)
r_s = bulk surface resistance (sm^{-1})
r_a = bulk aerodynamic resistance (sm^{-1})

Fortunately, it is not necessary for all of us to understand this equation because we can obtain the end result from other sources. The Meteorological Office use a modified version of this equation to calculate the seven-day potential evapotranspiration rate within each of 190 40 × 40 km square grids covering the whole of England, Scotland and Wales. As there is now a better understanding of the effect of crop height, etc. on the transpiration rate, the Meteorological Office have also refined the technique to take account of various crop types. The service has operated since 1978 and is known as MORECS (Meteorological Office Rainfall and Evaporation Calculation System). The following section describes how the data can be used to schedule irrigation.

Manual Water Balance Sheet

The evapotranspiration figure obtainable from the Meteorological Office can be used to construct a manual water balance sheet. This is a simple technique and there are several versions available, but all are similar in their approach. The technique has been used for many years now and has proved itself to be a successful and reliable method of planning irrigation on a day-to-day basis.

The concepts of *soil moisture deficit* (SMD) and *critical soil moisture deficit* (critical SMD) were described fully in earlier chapters. The principle behind the balance sheet is that the SMD is calculated on a daily basis, and the crop is irrigated before the critical SMD is reached (see Figure 4.1). The approach is based on the concept of a water balance. The water that enters the soil (rainfall and irrigation) is treated as a credit, and the water that exits (drainage, run-off and evapotranspiration) is recorded as a debit. The difference between the two is taken as representing the soil moisture status.

One of the earliest published accounts of a water balance sheet was contained in MAFF's *Technical Bulletin No. 4* in 1954. The technique was simple to use and the same basic concept is still used today. Balance

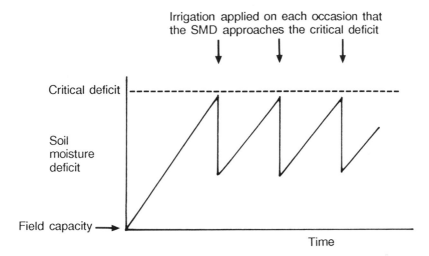

Irrigation applied on each occasion that
the SMD approaches the critical deficit

Critical deficit

Soil
moisture
deficit

Field capacity →

Time

*Figure 4.1   Diagrammatic representation of irrigation scheduling using a
water balance.*

sheets, however, are difficult to apply to crops with incomplete cover.
The original Penman figure for potential evapotranspiration is relevant only to crops with full cover, and a correction is required when
it is used for crops that exhibit partial cover. The corrections used
in the early days, however, lacked precision.

The accuracy was improved when Stanhill of the National Vegetable Research Station at Wellesbourne developed his graphical method
of correcting for incomplete cover. This method is still used today
at some research stations, but is too complex and time-consuming to
be acceptable to most growers.

Several years elapsed until, in 1977, a fairly straightforward system
of accounting for crop cover was published. This method, developed
by R. Adams and S. P. McClean of ADAS, is sometimes referred
to as CML1 because it was published, in its final form, as *Cambridge
Meteorological Leaflet No. 1*. Most people have simply referred to
it as the ADAS method.

Rainfall is recorded on the farm using a suitably sited rain gauge.
The location of the rain gauge requires attention, because siting too
close to a building, etc. will lead to faulty readings. Some growers
have avoided using a rain gauge and have preferred, for the sake of
convenience, to rely on rainfall data provided by a nearby Met. Office
weather station. This practice is *not* recommended, as rainfall can vary
greatly over quite small distances, especially during summer showers.
Instances of 25 mm rainfall occurring at one site, with another nearby
site receiving nothing or very little, are commonplace.

Irrigation can be measured on site using 'catch-cans' which, as their name suggests, are merely empty cans strategically placed in the field and used to record the amount of water falling in them. Most farmers probably rely on the combination of manufacturers' data and their own experience with the equipment to judge the amount of irrigation applied, but measuring it is far better. Machinery handbook rates of work cannot be relied on for accuracy.

As already mentioned, an *evapotranspiration* figure is available from the Met. Office. Evapotranspiration depends not only on the weather, however, but also on the green leaf area ground cover which the crop has reached. For convenience, three stages have been identified in this respect:

1. Before the crop attains 20 per cent ground cover   During this phase, transpiration rates are low and water loss is mainly by evaporation from the soil surface. This is assumed to be equivalent to 1.5 mm per day, and any influence of weather upon this is likely to be small and is ignored. Because crops at this stage usually have a very shallow rooting system, water can only be lost from the upper layers, and some allowance must be made for the fact that evaporation will slow down as the soil becomes drier. This is done by setting a maximum deficit for the soil, and assuming that evapotranspiration ceases when the SMD reaches that maximum. The maximum is set at 20 mm for light and medium soils, and 30 mm for heavier soils.

2. Between 20 and 75 per cent ground cover   During this phase, transpiration (loss of water through the plants) is considered to be a more important component of water loss. The Met. Office publish a list of average evapotranspiration figures for each of the fifty-two climatic areas of England and Wales. These figures are obtainable from HMSO in *The Agricultural Climate of England and Wales*, MAFF/ ADAS *Reference Book 435*. Examples of these are shown in Table 4.1, drawn from agroclimatic area 16. A grower should use the figures for his own agroclimatic area for crops in this second phase.

During periods of above average sunshine, transpiration is likely to be higher than average, but such periods are often associated with a dry soil surface and evaporation rates are lower than average. These two effects are assumed to balance and the average evapotranspiration rate is used in the calculation. During cloudy or wet periods, when transpiration rates are lower than average, the ground is usually damp and Adams and McClean argued that the subsequent evaporation from the wet soil tends to be higher than average. They argued that it was again reasonable to assume a balance of the two effects, and the average

Table 4.1 Average daily evapotranspiration rates for agroclimatic area 16

Month	Average evapotranspiration (mm/day)
April	1.9
May	2.7
June	3.1
July	3.1
August	2.3
September	1.5

Source: ADAS.

evapotranspiration figures could again be used. I have used this technique often myself, but I have found it preferable to make a downward correction during periods of low evapotranspiration, as described in the next paragraph. This is supported by an ADAS investigation, using neutron probes in a potato crop of low leaf cover, which showed that evapotranspiration rates during cool and cloudy weather were considerably below average.

3. 75 per cent ground cover and greater During this phase, water loss is almost entirely via transpiration, and greater account must be taken of variations in the weather.

It is no longer sufficiently accurate to use average evapotranspiration figures over a period of more than a week, or errors will accumulate. To avoid this, the grower must obtain a true evapotranspiration figure from the Met. Office every week and correct his calculations accordingly. The average figures are used to calculate the SMD each day of a seven-day period, and then the Met. Office figure is used to correct the SMD at the end of each seven-day period. The correction will be the difference between the Met. Office total for the previous seven days and the grower's own total of average values.

Calculating a daily balance of rainfall, irrigation and evapotranspiration, with a weekly correction where necessary, produces an estimate of the SMD. It is usual to assume an SMD of zero (field capacity) on 1 April, and calculate from that date. Whenever the incoming rainfall and irrigation exceed the SMD, the excess is considered to be lost as drainage or run-off, and the SMD is then taken as being equal to zero, i.e. the soil is assumed to return to field capacity. Even though it may be incorrect to assume a state of field capacity on 1 April, it is likely in most years that the soil will return to field capacity some time shortly after that date, and any error is effectively cancelled out.

The technique is simple, successful and reliable. It is doubtful if anything more sophisticated is necessary for most situations. The amount of work involved in producing balance sheets, however, does appear quite considerable and, not surprisingly, the technique has not proved too popular with farmers. There are some areas where, because of the enthusiasm and help offered by local ADAS advisers, the technique became commonly used, but the majority continued to ignore it. Most farmers found it complicated and were dissatisfied with the amount of calculation required, especially when they had a large number of fields to irrigate.

There are some situations where the technique shows some deficiencies. The method of accounting for incomplete crop cover may be simple, but it is not always accurate enough. In the stages where crop cover is below 75 per cent, and standard figures are used for evapotranspiration, errors will occur, but it must be remembered that this technique was primarily designed for grass and arable crops which pass quite rapidly through these stages and soon reach 75 per cent ground cover, e.g. maincrop potatoes, sugar beet, cereals. The errors involved are quite small for these crops and can be ignored. But many horticultural crops exhibit low ground cover for extensive periods of time and then the technique may lead to some large cumulative errors. A more accurate method is required in these situations, such as tensiometers, or sophisticated computer calculations as described in the next section.

In the preceding sections it was mentioned that point measurements, such as are obtained with neutron probes and tensiometers, can be difficult to interpret when unintentional variations occur in water application, as a result of wind drift, overlapping of irrigated areas, etc. The balance sheet technique has the advantage of not being subject to the errors resulting from non-representative point measurements. Where a non-uniform application is intended, as with trickle irrigation, the balance sheet approach can be the most erroneous. In this situation some form of direct measurement, such as tensiometers or neutron probes, is preferred.

COMPUTERISED BALANCE SHEETS

With the increasingly wide use of computers in recent years, it was a natural step to abandon the manual water balance technique and switch to a computerised version. Computerised balance sheets have the advantage of convenience, are popular with growers, and have also allowed scientists the opportunity to overcome some of the inaccuracies inherent in a manual system. The manual system was intentionally kept as simple as possible, which meant some cutting of corners. If a computer is doing the work, however, it is no longer necessary to keep the calculations simple. The introduction of the computer to irrigation scheduling has brought a much higher degree of accuracy to the calculation of water balances.

The first computerised irrigation scheduling service introduced into the UK was that operated by Fisons Fertilisers at Levington Research Station from 1977. Each week, from the first week of April, meteorological data are collected from a mixture of Meteorological Office and independent sites spread throughout the country. The rainfall and irrigation records are also collected from each farm via telephone, telex or telefax. The data are fed into the computer and the resulting advice on irrigation is sent to the farmer on the same day by first class post.

In the early 1980s, ADAS experimented with computerised scheduling systems that could be used with on-farm microcomputers. After several years' research, the ADAS workers made the important decision to move away from on-farm microcomputers and switch their attention to developing a system that is run by advisers on a central computer. This decision was made for several good reasons:

1. It allows an ADAS adviser to check and interpret the advice before it arrives at the farm.
2. The MAFF computer is linked to the Meteorological Office computer at Bracknell and this allows the use of up-to-date meteorological data on any day of the week that the service is required to operate.
3. It allows the weather forecasters at the Meteorological Office to input weather forecasts into the system. This is considered to be a major advance in irrigation scheduling, as expected future weather trends can be used to plan irrigation requirements in advance.

This system evolved into the service now known as *Irriguide*, and was introduced commercially in 1987. It was developed primarily by Spackman, of the Meteorological Office and on secondment to ADAS, but many ADAS workers have played a large part in its development into a successful commercial venture. Alongside the use of weather forecasts, there are several other notable features of this service that are of considerable advantage. One of these is that not only does the computer program analyse meteorological data, it also simultaneously models the growth of the crop. All water balance calculations are sensitive to the amount of ground cover established by the crop. This has always placed a burden on the farmer continually to examine the crop and monitor crop cover. Irriguide overcomes this by using temperaure data each week to estimate the development of crop cover. At the time of writing, farmers are still advised to check crop covers occasionally and any figures they provide are fed into the computer, but results to date indicate that the estimates given by Irriguide are generally satisfactory.

In a similar manner, Irriguide also estimates root growth throughout the season and uses this to calculate how much water is available to the crop at any particular time. It thus calculates a continually changing critical deficit and warns when a crop is likely to come under stress. While the basic program is now completed, Irriguide is still being modified in order to incorporate 'safety' features as their need becomes apparent. For example, in order to prevent the use of spurious rainfall data, Irriguide compares all farm rainfall data against that from nearby weather stations. If a large discrepancy is apparent, Irriguide will give a warning, notifying the need for investigation into the siting of the rain gauge, etc.

Several computerised irrigation scheduling methods have now been developed for use in Europe. In France, a system known as Irritel is used, and much irrigation in East Germany is now scheduled via computers. A similar method being developed in Italy is known as the RR model, and the Danish Centre and Agrometeorological Service are also developing a program.

In 1989, systems such as Irriguide probably represent the state of the art in irrigation scheduling but, as technology races ahead in leaps and bounds, one cannot help but wonder what system we will be using in another twenty years.

HOW MUCH WATER TO APPLY

All the irrigation scheduling techniques discussed so far are concerned mainly with determining the *best time* to apply irrigation. We have not yet addressed the question of *how much water* to apply on each occasion. Many farmers always apply 25 mm but, although this is often a convenient amount, there is nothing sacrosanct about it. Adjustment can lead to more efficient overall use of equipment.

A recent experiment at Gleadthorpe EHF was designed to test this point. Two varieties of potatoes were chosen for the experiment, namely Pentland Crown and the processing variety Record. Both varieties were irrigated each time the critical deficit of 35 mm was approached. One treatment consisted of applying 25 mm irrigation each time, and the other consisted of applying only 10 mm. Both treatments received the same *total* amount of water during the season, but the second treatment required a more frequent application of the smaller amount. At harvest, the yield and the quality, including the specific gravity of the Record, were identical for both irrigation treatments. There was no advantage or disadvantage to be gained from rigidly applying 25 mm on each occasion. The amount of water to be applied at each irrigation should be determined by the state of the soil and the logistics of applying the water.

The first principle is an obvious one but needs stating nevertheless. Do not apply more water than can be accepted by the soil. For instance, if irrigation is being applied at an SMD of 15 mm, there is no benefit to be gained from applying more than 15 mm water. A single application of 25 mm is unlikely to cause any harm, but if this were repeated many times throughout the season, the crop might suffer. In sandy soils, there is a high risk of nitrogen being leached beyond the reach of the roots, which could produce lower yields. On heavier soils, there is a risk of waterlogging or soil structure damage, and perhaps increased disease. So the amount of water applied on each occasion should never exceed the amount required to return the soil to field capacity. Indeed, if there is a possibility of significant rainfall during the following two or three days, it is preferable not to return even to field capacity, but to leave a small SMD remaining so that the natural rainfall can be absorbed and effectively utilised. This is why so many irrigation plans take the form of 12 mm at an SMD of 15 mm, 15 mm at 20 mm, or 25 mm at 35 mm, etc. They all leave a small margin for possible subsequent rainfall, and maximise the efficiency of the irrigation water without risking crop loss.

Another important point to consider when deciding how much water to apply is the size of the field and the type of irrigation equipment available. In the UK, hosereel machines are the commonest; the amount of water applied is regulated by the speed of travel of the machine. During a dry spell, it is crucial that the best use is made of each machine, and this often means ensuring that the machine finishes a run at a convenient time of day to change lanes and begin the next run immediately. For instance, consider a field of such length that a run delivering 25 mm will take 17 hours. Starting at 7 a.m. the run will complete at midnight, and it is unlikely that the farmer will find it convenient to change lanes at that time, so the machine may be left standing until the next morning. If, however, the machine were operated at a faster pace, applying less than 25 mm water, it may be possible to switch lanes at 7 p.m., allowing full use of equipment throughout the complete 24-hour period.

There is less flexibility in the maximum amount that can be applied on each occasion. In many situations, 25 mm represents the upper limit that should be applied. This is particularly true with a crop such as potatoes, which is usually grown in ridges. Applying amounts in excess of 25 mm is not recommended as it will result in run-off, even on light soils. Some crops, such as sugar beet, are associated with high critical deficits late in the season, and it is possible to increase the application rate to 37 mm (1½ inches). Some farmers apply up to 50 mm each time, but this is not recommended, especially as the UK climate involves the risk of heavy rainfall occurring a couple of days after the irrigation has been applied. It must be remembered that soils differ markedly in the rate at which water can infiltrate. While the rate is surprisingly high for dry soils, it soon falls off to an equilibrium as the soil becomes wetted, particularly with heavier textures. Typical infiltration rates are presented in Table 4.2.

Table 4.2 Equilibrium infiltration capacities associated with some common soil textures

Category	Equilibrium infiltration capacity range (mm/h)	Textures
Very high	Greater than 100	Coarse sands, sands, loamy coarse sands, loamy sands
High	20–100	Sandy loams, fine and very fine sandy loams, loamy fine sands and loamy very fine sands
Moderate	5–20	Loams, silt loams, silty loams, clay loams
Low	Less than 5	Clays, silty clays, sandy clays

Source: ADAS.

'WHOLE-FARM' SCHEDULING

If each farmer had a single field to irrigate, and the equipment to do the job fairly quickly, irrigation would be a simple procedure. Any of the previously described scheduling techniques could be used without too much trouble. However, in practice, each field often takes several days to irrigate and a farm consists of many fields which may all require irrigation simultaneously. This is where the management skills of both the farmer and the agronomist are required to produce the best results.

As an example, consider a large field of potatoes with a critical deficit of 55 mm, irrigated with a single hosereel that takes three days to cover the area. It would be unwise to wait until the SMD approached 55 mm and then start irrigating. The first part of the field would be irrigated satisfactorily, but the SMD would be likely to exceed 55 mm on the final part before irrigation was applied, resulting in a possible loss of yield. With an average transpiration rate of 3–4 mm per day in eastern England in the summer, it would be far better to start irrigating at a trigger deficit of around 45 mm, thereby guaranteeing that the *whole* field has received irrigation before the critical deficit is exceeded.

If a single hosereel has to cover two fields of this size, an even earlier start would be advisable, with the first application going on at an SMD of 35 mm, thus ensuring that both fields are irrigated before the critical deficit is exceeded.

This seems fairly obvious, and is relatively straightforward in the above two simple examples, but the complexity of the situation on many farms requires careful planning. Consider a typical example of three fields of potatoes, three fields of sugar beet and six of cereal crops, with two hosereels to cover the entire area. This is a situation that requires careful management to reduce to a minimum the number of occasions on which more than two fields require irrigating simultaneously.

A good approach is to consider the farm as two separate units, one for each hosereel, with an equal mix of crops in each unit. Thus, Unit A may consist of two fields of potatoes, one of sugar beet and three cereals, and Unit B the remaining field of potatoes, the other two sugar beet fields and the other three cereal fields. During May, only the cereal crops need to be considered for irrigation, and each hosereel has three to cover. One of these three should have a trigger deficit such that irrigation can be completed before the critical deficit is exceeded. The second should be triggered two or three days before this, depending on the size of the field and how long it takes to irrigate it. The third cereal should have a trigger deficit that is reached even earlier. In this way, it should be possible to irrigate all six cereal fields before the critical deficit is exceeded on any of them. As the end of May is approached, it is likely that potatoes and sugar beet will also have need of irrigation (see chapters 5 and 6 for details of when to start irrigating these crops). Irrigation of cereals is a low priority compared with potatoes and sugar beet (see chapter 12 for discussion on priorities), so the cereals will be largely ignored, and the irrigation will be used on the root crops. In the same way, these fields should have their trigger deficits chosen in such a manner that the demand for irrigation is staggered.

There is, of course, a limit as to how far this process can be extended. With particularly light soils, and the low critical deficits associated with them, it may not be possible to stagger all the fields on the farm without resorting to trigger deficits that are impracticable. Before irrigation can start on the first field, there needs to be an SMD high enough to make irrigation worthwhile (usually about 12 mm). In situations where it is not possible for the available equipment to cover the various fields on the farm as required, there are two choices:

1. Invest in more equipment.
2. Continue at the present level and clearly identify certain fields for priority treatment.

The first choice has been dealt with in the previous chapter, and the second choice is the subject of a later chapter.

POSSIBLE FUTURE DEVELOPMENTS

Infra-red thermometry

In recent years, researchers overseas have given a lot of attention to the use of infra-red thermometry as a means of assessing crop water stress. It has been known since the 1800s that plants suffering from moisture stress are warmer than plants with adequate water. There was, however, no practical tool for measuring plant temperature on a field scale. In the mid-1970s relatively inexpensive and reliable infrared thermometers became available, which encouraged researchers in the USA to study the concept in detail.

The general principle is based on the fact that when plants are healthy, they are at optimum temperature. To maintain this temperature, the stomata must be open, and transpiration must be proceeding at the potential rate. In order for water to evaporate, it must absorb heat from its surroundings, and this is known as the latent heat of evaporation. As evaporation proceeds, the surroundings are continually cooled. It is this cooling effect occurring during transpiration that prevents the leaf temperature rising unduly. As plants experience moisture stress, the stomata close, the actual rate of transpiration decreases and the leaf temperature rises above that of unstressed plants.

To apply the principle to a field crop, it is necessary to know whether the measured temperature represents a stress, or whether it is the optimum temperature. But how do we know the optimum temperature? It would presumably be possible to maintain on each farm a well-watered area to compare with other fields, but researchers have developed a better approach.

The technique involves measuring both leaf temperature and air temperature, and calculating the difference. For any particular set of weather conditions it is possible to estimate a temperature difference that represents nil stress, a temperature difference that represents extreme stress, and produce a stress scale between these two. The actual temperature difference measured in the field is then used to estimate the degree of stress being experienced by the crop. This sounds complicated but is, in fact, quite simple because instruments have been developed that are capable of executing all these functions and producing a stress index for each crop.

As a result of its simplicity, this technique is seen as a major advance in Australia and the USA, and researchers in some other countries (e.g. India) are also claiming that it is successful under their conditions. For UK application there are several problems, however. The technique is only reliable under conditions that include a high solar radiation level and a high potential evapotranspiration rate. Only if the skies are clear and the temperature is in the 70s Fahrenheit are we likely to have success. Conditions in the UK often involve frequent cloud cover, only moderate temperature and relatively high humidity, all of which make the technique difficult to use.

A second problem is that even under American conditions, the technique may not be sensitive enough for potatoes. Researchers in the USA claim that while the method is excellent for a range of crops, potatoes are an exception, and they only show an appreciable change in leaf temperature at severe water deficits. In the UK, potatoes are the most important crop on many farms where irrigation is practised (40 per cent of the irrigated land in 1987 was planted with potatoes). Much of this irrigation was applied at deficits below 15 mm in order to control the level of common scab within the crop (see chapter 5 for details). Infra-red thermometry is unlikely to be suitable for this particular requirement even if calibrations are eventually found for our climatic conditions.

The measurement of leaf and stem thickness

Researchers at the Institute of Horticultural Research, Wellesbourne, have been looking at the relationship between moisture stress and changes in the stem diameter or leaf thickness of plants. They discovered that, as plants become stressed by lack of moisture, there is a microscopic, but measurable, decrease in both stem diameter and leaf thickness. If using young plants, there are complications arising from the fact that the plants are growing, but any change in the thickness of older leaves of brussel sprouts, for instance, was found to be determined solely by differences in water stress.

It will be interesting to see if this generally proves to be the case for a range of crops. Nevertheless, it is unlikely that the technique will be used for field irrigation scheduling as it only indicates the need or otherwise for irrigation at the time the measurement is made; it cannot be used to forecast a need for irrigation several days in advance. This may be sufficient in a glasshouse where frequent measurements can be made and irrigation applied as the need occurs. In a field situation, using mobile rainguns, it is often necessary to know several days in advance of requirement, in order that the overall irrigation of the farm can be planned correctly, as described in the previous section.

Another technique involves carefully listening to the plants with amplification equipment. As tall plants (e.g. nursery stock) come under stress, the columns of water within the plant stem break up, and the 'popping' noise created by this can be identified with sensitive equipment and amplified. This technique, however, is in the early stages of development, and again has the problem of being difficult to use to forecast irrigation needs several days ahead.

There is no doubt that, as time proceeds and technology progresses, other techniques of irrigation scheduling will be invented. For the immediate future, however, it appears that the only reliable techniques will continue to be direct measurement of soil moisture with tensiometers or neutron probes and, for most farmers, the use of computer evapotranspiration models such as Irriguide.

REFERENCES AND FURTHER READING

ADAMS, R. J. and McCLEAN, S. P. (1978), 'Day-to-day estimation of irrigation need', *Cambridge Meteorological Leaflet No. 1*, internal ADAS paper, no longer available.

BAILEY, R. J. and SPACKMAN, E. (1988), 'Irriguide — the ADAS way', *Irrigation News 13*, 19–24.

BELL, J. P. (1976), 'Neutron probe practice', *Institute of Hydrology Report No. 19*.

CARR, M. K. V. (1984), 'Irrigation scheduling in the United Kingdom', *Proceedings of the North-Western European Irrigation Conference*, Billund, Denmark, 110–23.

DENT, D. L. and HAY, R. D. (1988), 'Tensiometers', *Irrigation News 13*, 33–43.

GREACEN, E. L. (1981), 'Soil water assessment by the neutron probe method', published by CSIRO, Australia, pp. 140.

HESS, T. M. and MATHIESON, I. K. (1988), 'Irrigation management services', *Irrigation News 13*, 25–32.

JACKSON, R. D. (1982), 'Canopy temperature and crop water stress', *Advances in Irrigation, Volume 1* (Academic Press, New York), 43–85.

MAFF (1967), 'Potential transpiration', *Technical Bulletin No. 16*.

MAFF (1982), 'Irrigation', *Reference Book 138*.

MCBURNEY, T. (1987), 'The use of sensors for monitoring plant water potential', *Proceedings of the 2nd North-Western European Irrigation Conference*, Silsoe, UK.

MCBURNEY, T. and COSTIGAN, P. A. (1988), 'Continuous measurement of plant water stress', *Acta Horticulturae*, 228, 227–34.

MOGENSEN, V. O., SVENDSEN, H., JENSEN, S. E. and JENSEN, H. E. (1987), 'Canopy temperature in relation to soil water content', *Proceedings of the 2nd North-Western European Irrigation Conference*, Silsoe, UK.

PAULSON, G. A. (1988), 'Hydro fertilisers irrigation service', *Irrigation News 13*, 10–14.

SIVAKUMAR, M. V. K. (1986), 'Canopy air temperature differentials, water use and yield of chickpea in a semi-arid environment', *Irrigation Science 7*, 149–58.

SMITH, L. P. (1984), 'The agricultural climate of England and Wales', MAFF/ADAS *Reference Book 435*.

SMITH, R. C. G., BARRS, H. D. and STEINER, J. L. (1986), 'Alternative models for predicting the foliage-air temperature difference of well-irrigated wheat under variable meteorological conditions', *Irrigation Science 7*, 225–36.

STARK, J. C. and WRIGHT, J. L. (1985), 'Relationship between foliage temperature and water stress in potatoes', *American Potato Journal 62 (2)*, 57–68.

THOMPSON, N., BARRIE, I. A. and AYLES, M. (1981), 'The Meteorological Office rainfall and evaporation calculation system: MORECS', *Meteorological Office Hydrological Memorandum No. 45*, pp. 69.

TURVILL, C. G. (1988), 'The ICI irrigation service', *Irrigation News 13*, 15–18.

Chapter 5

Irrigation of Potatoes

It is well established that in seasons when moisture supply from rainfall is limited, irrigation can improve the yield and quality of potatoes. To achieve the full benefit, the irrigation must be planned, taking account of the crop, soil and weather conditions. In this chapter, the results from ADAS experiments will be described, showing the benefits that have been achieved from using water effectively on the potato crop, and also showing how irrigation should be planned.

YIELD RESPONSE

Early potatoes

The irrigation response of potatoes lifted in June has been investigated in Pembroke (1968–71) and at Rosewarne EHS (1973–5, 1981–5). The results are shown in Table 5.1. Irrigation increased yields in nine of the twelve years, albeit sometimes only marginally, with an average increase of 2.5 t/ha ware yield for an average water application of 55 mm. The very high response in 1982 is rather unusual, and reflects the low rainfall during spring and early summer of that year.

The need for irrigation obviously varied considerably from year to year, but dry spells commonly occur in May or early June which, if combined with high transpiration rates as a result of warm temperatures or windy days, cause the crop to come under moisture stress at a time when it should be bulking rapidly. The results show that irrigation can typically be expected to produce a yield response of 1.5–2.0 t/ha in an average year from a lift in mid-June.

These experiments also provide guidelines on how to plan the irrigation of early potatoes. Table 5.2 shows the yield responses obtained at Pembroke with two different irrigation schedules, viz. 18 mm water applied when the SMD reached 18 mm, and 38 mm water applied when the SMD reached 38 mm. The first treatment was clearly superior, showing that a deficit of 38 mm is too high for early potatoes on light

Table 5.1   Yield response from irrigation of early potatoes in ADAS experiments. Each result is a mean of several lifts throughout June

Year	Site	Irrigation applied (mm)	Yield response (t/ha)
1968	Pembroke	25	−0.4
1969	,,	0	0
1970	,,	50	5.5
1971	,,	45	1.2
1973	Rosewarne	75	1.7
1974	,,	50	2.1
1975	,,	100	3.2
1981	,,	0	0
1982	,,	100	11.0
1983	,,	100	2.4
1984	,,	75	1.6
1985	,,	50	1.1

Source: ADAS.

soils. The more recent experiments at Rosewarne EHS also investigated different irrigation plans and the results are shown in Table 5.3. The plans were as follows: (a) 25 mm water every 10–14 days after 100 per cent emergence, regardless of weather conditions; (b) 25 mm water applied every time the SMD reached 25 mm regardless of crop development; (c) 25 mm water applied every time the SMD reached 25 mm, but only after the tubers had reached marble size. The results show that irrigating every 10–14 days regardless of weather conditions was clearly inefficient and gave the lowest yield response. There was little overall difference between the other two treatments, but in the driest years there was an advantage from irrigating as soon as the deficit reached 25 mm, rather than waiting until the tubers reached marble size. For practical reasons this is also advisable, as it allows more time to move the irrigation system over the entire crop.

Table 5.2   Effect of two different irrigation plans on yield response of early potatoes in Pembroke

Year	Yield response (t/ha) Irrigation plan	
	18 mm water applied at 18 mm SMD	38 mm water applied at 38 mm SMD
1970	5.5	3.6
1971	1.2	0.8

Source: ADAS.

Table 5.3   Effect of various irrigation plans on yield response of
early potatoes at Rosewarne EHS

	Yield response (t/ha) Irrigation plan		
	25 mm applied	25 mm at	25 mm at
	every 10–14 days	25 mm SMD	25 mm SMD
Year	from emergence	from planting	from marble size
1982	9.6	11.0	8.5
1983	−4.1	3.5	1.5
1984	−0.5	1.0	1.6
1985	0.7	0.8	1.0
Mean	1.4	4.1	3.2

Source: ADAS.

Second early potatoes

Experiments were carried out at Gleadthorpe EHF between 1960 and
1968 on crops lifted in early July. The results are shown in Table 5.4.
Irrigation increased yields in eight of the nine years, with an average
increase of 4.4 t/ha ware yield for an average water application of
70 mm.

Table 5.4   Yield response from irrigation of second early potatoes
at Gleadthorpe EHF

Year	Irrigation applied (mm)	Yield response (t/ha)
1960	102	6.3
1961	97	5.0
1962	81	5.0
1963	53	2.7
1964	66	2.5
1965	48	0.5
1966	51	5.1
1967	112	9.1
1968	25	3.5

Source: Gleadthorpe EHF, ADAS.

In some years, the irrigated and unirrigated crops were lifted on
a series of dates, starting in late June. In these years, when the crops
were lifted in late June, the yield increases given by irrigation were
less than shown here, but were greater with mid- to late July liftings.

It is difficult to translate this yield response into a financial benefit. Using an average price per tonne for first or second early potatoes and applying it to the yield responses described here is likely to lead to an underestimate of their financial value, as those years which show a good response to irrigation are usually associated with higher than average prices.

Maincrop potatoes

The response of maincrop potatoes to irrigation has been studied in experiments at Gleadthorpe for thirty years since 1958. In fact, 1975 is the only year without data, because a freak hailstorm damaged the experiment in that year. The results, averaged over many varieties, are shown in Table 5.5 in order to demonstrate the enormous variation that exists between years with respect to both yield response and water application.

The data are summarised into a more manageable form in Table 5.6. The thirty years have been split into three equal-sized groups by collecting the ten lowest response years together, the ten middle and the ten highest. Over the whole thirty years the average yield increase was 10.6 t/ha with a very large contribution from the ten highest response years. Indeed, the contribution from these ten years amounted to 75 per cent of the total yield response obtained over the complete thirty-year period.

As with early potatoes, it is in just such years that the price for potatoes is above the average, so the contribution of these years to the gross cash return from irrigation is likely to be considerably more than 75 per cent. It follows that the capacity of an irrigation system to water adequately in these years is of great importance. This capacity may be limited by the amount of abstraction allowed by the water authority, or because of reservoir size, or a lack of irrigation machinery or labour.

In order to achieve the level of yield increase given in dry years in the Gleadthorpe trials, a farmer would need to install equipment capable of putting on the amount of water applied in the trials — and this was anything up to 275 mm (1976)! As this would be impractical, it is useful here to estimate the yield responses that would have been obtained if the amount of water available was limited to the levels commonly found in farm practice. It is not possible to do this precisely because the experiments did not always include a treatment that represented 'limited irrigation'. However, it is possible to obtain an approximate estimate. In wet years, when little irrigation was applied, the yield response obtained in the trial can be taken as being equal

Table 5.5 Yield response from irrigation of maincrop potatoes at
Gleadthorpe EHF

Year	Irrigation applied (mm)	Yield response (t/ha)
1958	51	−1.5
1959	203	25.6
1960	150	7.3
1961	142	3.8
1962	132	3.3
1963	43	0.8
1964	86	2.8
1965	41	−1.0
1966	99	1.3
1967	152	16.1
1968	51	0.5
1969	102	14.8
1970	226	24.8
1971	84	3.5
1972	99	12.6
1973	64	0.8
1974	79	−1.3
1975	No data	No data
1976	272	39.7
1977	111	23.0
1978	91	3.9
1979	120	18.9
1980	66	−0.8
1981	113	25.3
1982	115	8.6
1983	165	23.0
1984	137	23.9
1985	143	13.1
1986	154	17.9
1987	83	3.8
1988	142	2.9

Source: Gleadthorpe EHF, ADAS.

to that obtained in practice even if irrigation were limited. In dry years, when high amounts were applied, it is necessary to make allowance for limited irrigation and arbitrarily adjust the response downwards. I have assumed a linear response in dry years, i.e. if only half the required water is applied in a dry year, only half of the potential yield response would be obtained in that year. This assumption will not always be valid, especially in wet years but, for the purposes of this exercise, it should be sufficiently accurate to estimate responses to limited irrigation in dry years.

Table 5.6 Irrigation of maincrop potatoes at Gleadthorpe EHF, 1958–88

	Ware yield increase over unirrigated potatoes (t/ha)
Lowest response (10 years)	0.5
Medium response (10 years)	7.5
Highest response (10 years)	23.8
30-year average	10.6

Source: Gleadthorpe EHF, ADAS.

Using this technique it is possible to produce an estimate of yield response at Gleadthorpe over the thirty-year period for a range of irrigation capacities (see Table 5.7). The figures demonstrate the effect of limiting irrigation in a dry year, particularly if the application capability is less than 150 mm. It must also be remembered that limited or insufficient irrigation will often fail to produce the high quality skin-finish that is desired by most growers today (see section beginning on p. 70 for a more detailed account of irrigation and potato quality).

Table 5.7 Estimate of maincrop potato irrigation responses at various levels of irrigation capacity at Gleadthorpe EHF

Maximum amount of irrigation available in a dry year (mm)	Average amount used (mm)	Estimated average response (30 years) (t/ha)
100	88	7.0
150	108	9.4
200	114	10.1
250	116	10.5
275	117	10.6

In summary, the yield response in dry years is around 24 t/ha on light soils, but on average the response over a number of years is 10–11 t/ha, or somewhat less if irrigation capacity is limited. On heavier soils, with a greater available water capacity, a smaller response can be expected. For example, it has been estimated that potatoes on the medium silt soils in the dry eastern counties are likely to show an average response of 5 t/ha. Specific information can be obtained for any particular site by analysing the soil and local weather patterns, as described in chapter 3.

Varietal variation

The yield responses quoted so far are averaged over many varieties, but experiments at Gleadthorpe EHF have shown that varieties differ greatly in their response to irrigation. This can be very useful in choosing priorities for irrigation if the system has insufficient capacity to cope with all the potatoes on the farm. Further details can be found in chapter 12.

An irrigation plan for maximising the yield of maincrop potatoes

As explained in chapter 4, two common methods are in use for determining when a crop requires irrigation. These involve either the *calculation* of the soil moisture deficit (SMD) or the *measurement* of soil moisture with instruments. During dry periods the SMD increases and, if it exceeds a certain level, yield is likely to be affected. This level is called the critical deficit. It will vary according to soil texture and rooting depth. Examples of critical deficits for potatoes are shown in Table 5.8. Irrigation should be planned to keep the SMD below the critical deficit at all times. As an example, on a loamy medium sand over sand, a typical irrigation plan would apply 25 mm whenever the SMD approached 35 mm. These critical deficits are only guidelines. For a full description of how to calculate critical deficits for different crops and different soil types, see chapter 2.

The alternative technique for irrigation scheduling involves tensiometers. Where tensiometers are used, the soil moisture tension should be kept below 30 centibars.

When to stop irrigating

The choice of date for the final application of water to a potato crop is a compromise. There is a yield penalty associated with stopping too early, because the crop may come under drought stress before it matures. But there may be a worse penalty if irrigation is continued too late, as wet soil conditions at harvest lead to many difficulties, including disease in store. To complicate matters further, the use of a desiccant may be dependent on the soil not being too dry at the time of application.

Table 5.8 Critical soil moisture deficits for maincrop potatoes according to soil texture (assuming a rooting depth of 70 cm)

Soil texture	Critical SMD (mm)
Sand	30
Loamy medium sand over sand	35
Loamy fine sand over sand	40
Medium sandy loam	55
Sandy clay or sandy clay loam	55
Clay or silty clay	55
Clay loam or silty clay loam	60
Fine sandy loam	60
Sandy silt loam	60
Silt loam	75
Acid or shallow peats	85*
Deep fen peats	120*

Note: * Although organic soils have high critical deficits, irrigation is normally applied earlier in order to avoid excessive drying of the ridge and subsequent problems of rewetting the soil.

I find the following guidelines quite useful. The time to stop irrigating is dependent on the AWC of the soil.

1. On lighter soils where the AWC of the profile is 10 per cent or less and drainage is not a problem, harvesting difficulties are rare and irrigation can continue late into the season. In practice, growers usually continue until the end of August, provided that the crop is still green.
2. On medium soils, with an AWC of 15 per cent, the middle of August is more suitable.
3. On heavy soils, where the AWC of the profile is 20 per cent or more, harvesting conditions merit greater consideration, and irrigation should be stopped well before maturity, *usually* at the beginning of August. Exceptionally, irrigation can continue into the second week of the month if there is a danger of significantly exceeding the critical deficit *and dry weather is forecast.*

There are circumstances, however, when irrigation should stop even earlier.

1. If the crop is already maturing, irrigation should be stopped as soon as the green leaf cover is reduced to about 80 per cent. This should ensure that transpiration levels are still high enough to use up most of the applied water before harvesting is started.

2. At least two weeks should be allowed between the final irrigation and desiccation of the crop, otherwise the skins may not set satisfactorily. This is particularly important for crops destined for processing, as dry matter levels may be reduced if the soil is still wet when the crop is desiccated.
3. Once a crop develops obvious blight in the foliage, it should not be irrigated as this increases the risk of carrying the infection down to the tubers.
4. Earlies and second earlies destined for 'green-top' lifting may be irrigated up until three days before starting to lift on light soils, provided the weather forecast is fine. On medium or heavier soils, however, irrigation should be stopped at least a week before the planned date of lifting.

Common Scab Control

Yield is not the only criterion for success in potato production. Quality and marketability of produce are also very important. Experiments have shown that irrigation produces a marked improvement in quality.

Common scab is caused by several closely related soil bacteria, usually grouped under the name *Streptomyces scabies*, and is one of the most widespread diseases affecting the potato tuber. The disease causes serious loss of skin quality in potatoes, particularly in susceptible varieties. Research has shown that tubers are attacked as new tissue is forming. Young skin over new tissue contains small pores called stomata, but as the skin matures these develop into larger pores called lenticels. The bacteria responsible for common scab can infect tubers through stomata and newly forming lenticels, but not through mature lenticels. The new tissue is susceptible to attack for a period of ten to fifteen days only, and it then becomes immune for the rest of the season. As a tuber grows, fresh areas of new tissue at the rose end are vulnerable, but the pattern of growth is such that *most* of the transition from stomata to lenticels is completed during the first six weeks after tuber initiation. The major concern, therefore, is to control common scab during the initial six weeks.

Control of this disease is based upon the fact that moisture on the tuber surface prevents infection by the bacteria. Work at Rothamsted Experimental Station and elsewhere has shown that tubers are only infected if the soil is dry as the new tissue is forming. Between 1969 and 1971 trials at Gleadthorpe EHF, in co-operation with Rothamsted, showed how irrigation can be used to control the disease. Watering whenever the SMD reached 15 mm for periods varying between two and six weeks after the ends of the stolons began to swell (tuber initiation) was compared with watering at 38 mm SMD or no irrigation, using the varieties King Edward, Majestic, Pentland Crown and Record. The effect on King Edward and Majestic is shown in Table 5.9.

Table 5.9 Effect of watering frequency on common scab infection (mean percentage of tuber surface affected at maturity)

		At 15 mm SMD for 4 or 6 weeks then at 38 mm SMD	At 15 mm SMD for 2 or 3 weeks then at 38 mm SMD	At 38 mm SMD throughout	No water
King Edward	1969	0.6	0.7	3.0	8.4
	1970	1.5	4.2	5.3	19.4
	1971	2.0	3.3	5.7	31.4
	3-year means	1.4	2.7	4.7	19.7
Majestic	1969	1.6	3.8	8.1	25.6
	1970	2.8	7.1	9.0	32.3
	1971	3.5	5.9	10.4	29.9
	3-year means	2.6	5.6	9.2	29.3

Source: Gleadthorpe EHF, ADAS.

Applying water whenever the SMD reached 15 mm for four to six weeks resulted in very low levels of scab infection, but unirrigated tubers were badly infected. Irrigating at 15 mm SMD for two or three weeks or whenever SMD reached 38 mm was insufficient to give clean tubers.

Further trials in 1982–4 showed similar results (Table 5.10) although the level of control was not as good.

The results show that common scab can be partially controlled by keeping the soil near to field capacity for at least four weeks after

Table 5.10 Incidence of common scab (% surface area) under different irrigation treatments at Gleadthorpe EHF, 1982–4

Year	Variety	SMD at which irrigation was applied (mm)				
		15	35	45	55	Nil irrigation
1982	Pentland Javelin	2.0	2.1	6.1	+	4.8
1983	Estima	9.2	24.7	26.3	39.4	33.3
1983	Pentland Crown	10.2	16.6	14.8	34.2	28.8
1984	Maris Piper	34.0	+	+	+	66.9

Source: Gleadthorpe EHF, ADAS.
Note: + Indicates that treatment was not included in the experiment.

tubers begin to form (see Colour plates 9 and 10). Three theories have been put forward to explain this: (a) wet soil may promote disease resistance in the tuber, for example, by accelerating the rate at which stomata form into resistant lenticels; (b) a direct effect on the pathogen as a result of lower temperature or lower oxygen availability; (c) the bacteria responsible for the disease may be antagonised by other microbes. The concept of microbial antagonism has received most attention in recent years. In 1962, Lewis at the University of Nottingham reported that other bacteria were more frequent and common scab bacteria less frequent on tuber surfaces in wet than dry soil, and suggested that antagonism by other bacteria may account for the lower levels of disease found in wet soil.

A thorough study of this was undertaken by Adams and Lapwood at Rothamsted Experimental Station in the 1970s. They concluded that the scab control achieved in wet soil was probably caused by some form of microbial antagonism. The rate at which lenticels formed from stomata was unaffected by soil moisture, so the explanation of wet soil affecting lenticel development seems unlikely. These workers also drew attention to the fact that soil moisture appeared not to affect the pathogen directly, since it grew well in wet sterile soil. Furthermore, American scientists have shown that it can infect tubers under wet sterile conditions. Adams and Lapwood also found similar results to Lewis, namely that soil moisture resulted in a far wider spread of other bacteria over the tuber surface, and less common scab bacteria. Also, importantly, they showed that this effect on soil microflora occurred on that part of the tuber containing stomata and thus known to be susceptible to common scab. It does seem, therefore, that some form of microbial antagonism is the most likely explanation for irrigation helping to control common scab. However, it is still unknown whether the antagonistic bacteria are producing antibiotics which inhibit the scab organism, or whether they are merely better competitors in wet soils, where oxygen supply will normally be restricted.

Using irrigation to control this disease requires an intensive irrigation plan during the 4–6 week period (6 weeks is far safer) immediately following tuber initiation. It is not possible to state precisely the critical deficit for common scab control because there have been no experiments to measure the levels of common scab over a comprehensive range of soil moisture deficits, e.g. 10, 12, 15, 18 mm, etc. However, the experiments have shown a consistently good result from keeping the SMD below 15 mm, and this is advised where possible. Where tensiometers are used, the soil moisture tension should be kept below 20 centibars throughout the period. Rainguns are used successfully in commercial crops but this often results in the SMD exceeding 15 mm within part of the crop. Usually, irrigation is started with 12 mm water as the SMD reaches 12 mm, and the cycle continues over 3 days. Assuming a typical transpiration rate of 3 mm per day, the SMD passes through 15 mm on the second day and eventually reaches 18 mm on the third day as irrigation is applied to the final part of the crop. Attempts with 4-day cycles (or longer) inevitably result in the SMD exceeding 18 mm at some stage, with an increased risk of common scab infection, unless the weather is dull enough to keep transpiration rates well below 3 mm per day.

It would appear reasonable to assume that less intensive irrigation is required for common scab control on soils with more available water than the sands at Gleadthorpe. Unfortunately, however, there is no experimental data at present to test this view. In 1988, several commercial growers on fine sandy loam/silt loams, under ADAS supervision, used a regime which consisted of applying 15 mm whenever the SMD reached 20 mm for six weeks and achieved satisfactory results. Current ADAS experiments are designed to examine the intensity of irrigation required for common scab control on heavier soils. Until these experiments are complete, it is still advisable to follow the intensive schedule described above in all situations where common scab can be a problem.

It is crucial to start irrigating early enough if common scab control is required. In the Gleadthorpe experiments, irrigation was started at tuber initiation and it has now become common practice to withold the first irrigation until the crop has reached tuber initiation. It is conceivable that this practice may sometimes result in imperfect control for three possible reasons:

1. In some years the weather may be drier than those years in which the experiments were conducted. This could lead to high SMDs occurring before tuber initiation, creating very dry conditions in the ridges, which may subsequently become difficult to wet properly.
2. Commercial farms commonly use rainguns and this may result in less thorough rewetting of the ridge if it has been allowed to dry out.
3. Many growers wait for tuber initiation but miss it, because of the difficulty of determining the growth stage precisely. They wait until tubers are found on the longer stolons near the outside of the ridge, or even until flowering, before starting irrigation. *This is too late.* The first stolons to initiate tubers are the shorter ones in the centre of the ridge and the onset of flowering has *no* relationship to tuber initiation.

Scientists at the Scottish Crop Research Institute have studied the relationship between emergence and tuber initiation (see Table 5.11). They showed that tuber initiation occurred two to three weeks after emergence. Thus, to achieve a high level of control of common scab, it may be better to apply the first irrigation starting ten days after crop emergence, rather than delay until tuber initiation is observed, assuming, of course, there is a soil moisture deficit. Unfortunately, however, this may lead to a higher risk of powdery scab (*Spongospora subterranea*) and blackleg (*Erwinia carotovora* var. *atroseptica*) in some circumstances (see following section on other diseases). Growers are therefore still advised to withhold irrigation until tuber initiation, unless they are irrigating a variety with a high level of resistance to both these diseases.

Table 5.11 Time of tuber initiation (days after emergence)

	1981	*1982*	*1983*	*1984*
Maris Piper	21	16	20	16
Pentland Dell	20	18	—	—
Guardian	18	13	18	—
Maris Bard	—	—	13	—

Source: Scottish Crop Research Institute.

IRRIGATION AND OTHER POTATO DISEASES

The effect of irrigation on common scab has been described in detail in this chapter, but there are several other diseases that are affected in some way by irrigation.

Powdery scab

Powdery scab is caused by the fungus *Spongospora subterranea*. Unlike common scab, it is favoured by wet soils and the wetter than average summers of 1985–8 were associated with a considerable increase in the frequency with which the problem was reported. In its severest form the disease can make the crop completely unmarketable (see Colour plate 11).

Experimental work at the North of Scotland College of Agriculture has shown that powdery scab is often associated with restricted drainage and poor cultivation practice. It is essential that growers pay particular attention to soil management. For example, stone separation should not be carried out when the soil is still wet at depth, or a pan will be created which may subsequently prevent irrigation water or rain from draining through adequately, and create partial waterlogging.

As the disease is associated with wet soil conditions it might be expected to be particularly bad where irrigation is practised, especially the intensive irrigation required by common scab control. In 1982, a joint Rothamsted/Gleadthorpe EHF investigation involved planting infected seed at Gleadthorpe, and the effect of different irrigation regimes on the progress of the disease was studied (Table 5.12).

Table 5.12   Incidence of powdery scab (% tubers) under different irrigation regimes

| Year | Variety | SMD used to trigger irrigation | | | | |
		15	35	45	55	None
1982	Pentland Javelin	2.7	4.2	2.1	+	4.2
1983	Estima	9.7	1.3	2.6	0.7	0.0
1983	Pentland Crown	5.4	1.7	1.4	1.8	1.4
1984	Maris Piper	27.0	+	+	+	3.0

Source: Gleadthorpe EHF, ADAS.
Note: + Treatment not tested.

Although there was no effect in 1982, in the two years 1983 and 1984 intensive irrigation as practised for common scab control did increase the levels of powdery scab. This demonstrates that there is a potential problem where severely infected seed is planted and the crop then irrigated to control common scab.

Researchers in Australia have shown that irrigation before tuber initiation (a common practice in that part of the world) could lead to a much higher risk of powdery scab. Their site had a previous history of five successive years of cropping with potatoes, with powdery scab occurring each season. Delaying irrigation until tuber initiation or three weeks later significantly increased the proportion of tubers free of powdery scab (see Table 5.13).

Table 5.13   Effect of irrigation treatments on the severity of powdery scab (Australia)

Irrigation treatment	Tubers free from powdery scab (%)
Irrigated from planting	78.5
Delayed until tuber initiation	93.4
Delayed until tuber initiation + 3 weeks	95.6

Source: Irrigation Research Institute, Tatura, Victoria.

A similar experiment conducted in a greenhouse, using infested soil mixed with infested potato debris, showed that delaying irrigation until three weeks after tuber initiation was the best treatment where there were high levels of soil infestation.

What advice can be given to UK growers regarding the risk from irrigation if seed is healthy, but there is a history of powdery scab in the field? In this situation, irrigation may bring an unacceptable risk of further infection, but this risk is difficult to evaluate.

The areas of powdery scab infestation at Gleadthorpe, resulting from the experiments described here, were tested again after four years by using a susceptible variety (Estima) and intensive irrigation, but powdery scab did not show itself again. However, there have been reports from commercial farm practice suggesting that irrigation may be unwise where there is a history of powdery scab. If it is used, great care must be taken to avoid ponding. At the very least a resistant variety should be used, in conjunction, where possible, with a rotation of at least six years.

Where seed stocks are healthy and there is no history of the disease, the risk from irrigation cannot be totally discounted, but it is small compared with the risk of low yields and poor quality resulting from a drought. Many crops of potatoes, including the Gleadthorpe commercial crops, are irrigated intensively and only show low and sporadic levels of this disease. Where there is no history of powdery scab, the use of healthy seed, good cultivation practice and a rotation of at least four years, along with careful irrigation scheduling, should keep the risk of powdery scab to a minimum.

Potato blight

The greatest need for irrigation naturally occurs in the dry and sunny periods that are not associated with blight spread. However, it is still usual to continue to apply some irrigation even in dull, damp weather, as rainfall during such periods is often still inadequate. If a crop is irrigated, it must be considered as being in a 'high-risk' situation for potato blight (*Phytophthora infestans*), and disease control should be planned accordingly.

The best protection for foliage should result from the continual presence of an adequate level of fungicide. Rainfast fungicides and those with effective systemic or partially systemic components are likely to be most effective. All fungicides should be applied according to the manufacturers' recommendations, after irrigation whenever possible, to protect the crop until the next irrigation. This is not always easy. On heavier soils it may not be possible to travel shortly after irrigation. Some compromise is therefore necessary; the fungicide must be applied whenever required, ensuring that there is enough time for any systemic component to become absorbed before the next irrigation. Where control has failed and crops have obvious blight in the foliage, they should not be irrigated as this will increase the risk of carrying the infection down to the tubers.

Blackleg

Overwatering will create conditions favourable for the development of the bacterial disease blackleg (*Erwinia carotovora* var. *atroseptica*, see Colour plate 12). The bacteria enter the crop by various means including groundkeepers and aphids, but by far the most important source of inoculum is the seed tuber. If seed coated with *Erwinia* is planted into wet soil, the disease can rot the seed tubers and completely kill the plant before emergence. Sometimes the seed tuber does not rot completely and the plant emerges, but infection spreading from the base of the stem gives rise to the classic 'blackleg' symptoms. The risk of this is also greater in wet soils, because wet conditions favour the infection process.

Very wet soil conditions can also increase the spread of the disease, because the bacteria are carried through the soil by water, especially where watering is excessive and there is much movement of water through the ridge. The bacteria can be carried from rotting seed tubers to the daughter tubers or, in extreme cases, to daughter tubers of nearby plants.

Wet conditions late in the season can also result in the harvested tubers being smeared with the bacteria, creating a severe problem during subsequent storage.

A good irrigation scheduling technique will reduce these risks. Application of the correct amount of water will reduce the risk of disease development caused by wet conditions. It will also reduce the spread of the disease, as there will be less lateral spread of water throughout the soil. Finally, making the correct decision when to stop irrigating can reduce the risk of wet soil conditions late in the season, and the spread of bacteria on the surface of the tubers at harvest.

Other diseases

Stem canker (*Rhizoctonia*) causes most damage to plants affected by drought stress. Irrigation will help the crop overcome the worst effects of stem canker.

Finally, in a recent study by Hide at Rothamsted, irrigation decreased the level of silver scurf (*Helminthosporium solani*) on tubers. In the same study, skin spot (*Polyscytalum pustulans*) was found to be increased by irrigation, while no consistent effects were found on black scurf (*Rhizoctonia solani*).

IRRIGATION AND OTHER ASPECTS OF TUBER QUALITY

Growth cracking and second growth

It has been suggested that tuber growth cracking is a result of rapid changes in tuber growth rate, the causes of which are not fully understood. Some workers concluded that growth cracking was associated with mid-season water stress followed by a resumption of water supply. Others suggested that the prerequisite for growth cracking was any set of conditions which produced very rapid growth of tubers, which may or may not involve water stress. R. Jeffries and D. Mackerron at the Scottish Crop Research Institute (SCRI) have studied the problem in recent years and have suggested that much of the confusion is due to the fact that there are different forms of growth crack, induced by different conditions. These workers showed that conditions of very rapid growth, which in their experiment led to a 54 per cent yield increase in one week, resulted in very large cracks in the tubers of the variety Guardian (see Colour plate 13). This was the case even though the SMD prior to this period had never exceeded 20 mm, and moisture stress could not have been involved. Conversely, a shallow but extensive pattern of cracking occurred on the variety Record when water was applied after the SMD had reached 70 mm. In this latter case, the amount of cracking was greater when rewetting was greater, and also when rewetting occurred from greater deficits (see Colour plate 14).

The actual mechanism of both types of growth cracking is still largely unexplained, as is the fact that varieties differ in their susceptibility to both types. To add to this confusion, work at Gleadthorpe has shown that the form of cracking pictured in Colour plate 13 can also be related to soil moisture. It is clear that water stress plays a part and we can sometimes reduce the incidence of cracking by applying irrigation. A farm crop of Record which was monitored during the SCRI farm survey of 1984 was irrigated on one part of the field, and the other left unirrigated. In the irrigated part there was no growth cracking, but in the unirrigated part 50 per cent of the ware-sized tubers had growth cracks.

An experiment at Gleadthorpe in 1986 provided another clear demonstration of this effect of irrigation. Four varieties were each grown under two irrigation regimes: (*a*) an SMD kept below 35 mm at all times with irrigation; (*b*) unirrigated until the SMD reached 62 mm in early July and then irrigated as per the other treatment. The results (shown in Table 5.14) show that proper irrigation throughout the season markedly reduced the level of growth cracking in a susceptible variety such as Estima.

Table 5.14 The effect of irrigation upon the number of tubers showing growth cracks (thousands/ha) at Gleadthorpe EHF, 1986

Variety	SMD kept below 35 mm throughout the season	SMD allowed to rise to 62 mm before irrigation applied
Estima	7.9	21.0
Record	7.0	10.7
Cara	0.6	1.2
Pentland Dell	0.2	0.2

Source: Gleadthorpe EHF, ADAS.

The formation of 'dolls' appears to be a related phenomenon to growth cracking, and can be induced by moisture stress during tuber growth followed by a resumption of water supply (see Colour plates 15 and 16). The narrowness of the constriction is an indication of the severity of the moisture stress which, in an excessive case, may result in the complete cessation of tuber growth while the stolon apex continues to grow. The stress-sprouts which are then formed can reinitiate tubers when the soil is rewetted, and this will result in chained tubers. Proper use of irrigation, preventing the build-up of large SMDs during the season, will reduce the incidence of such problems.

Tuber dry matter

The concentration of dry matter in tubers is a major concern when crops are grown for processing. Both the crisping and chipping industries require potatoes with a high dry matter. This is normally measured as tuber specific gravity. The minimum acceptable limit for specific gravity of crisping potatoes is generally 1.080 or 20 per cent dry matter, but the ideal is considered to be 23 per cent dry matter. For french fry production, the minimum specific gravity is 1.075 or 19 per cent dry matter. Low dry matter potatoes cost more to fry, absorb more oil and give a less crisp product.

It is well known that potatoes grown in wet seasons tend to have low dry matter. This is not simply an effect of the amount of rainfall, however, but is largely the result of the cool temperatures and lack of sunshine associated with wet seasons. Of much more relevance to the grower is whether irrigation by itself can lead to changes in dry matter concentration.

Experiments have shown that irrigated crops do have lower dry matter concentrations than droughted ones but the difference is not great, except in years when large amounts of irrigation are applied. Of course, such years are typically very sunny and tuber dry matters tend to be high anyway, so the effect of irrigation is unlikely to cause the crop to fall below the threshold (see Table 5.15).

Table 5.15   Tuber dry matter (%)

Year	Variety	Unirrigated	Irrigated	Decrease
1983	Pentland Crown	24.6	23.6	1.0
	Pentland Dell	28.2	25.5	2.7
	Desiree	25.7	24.3	1.4
	Maris Piper	28.3	25.7	2.6
	Record	29.1	27.5	1.6
1984	Desiree	23.0	21.0	2.0
	Maris Piper	27.7	24.0	3.7
	Record	29.6	25.3	4.3

Source: Scottish Crop Research Institute.

Reducing sugars

A reducing sugar level of less than 0.2 per cent is required by the processing industry to produce the pale coloured crisps which are preferred by the majority of consumers. In ADAS experiments, irrigation has not affected tuber reducing sugar content or crisp colour, either at harvest time or after storage.

Storage potential of the crop

In a collaborative series of trials between Gleadthorpe EHF and the Potato Marketing Board, the keeping qualities of irrigated and non-irrigated crops were examined for seven years. In these trials, the irrigation was scheduled and applied correctly. There was not one instance of irrigation increasing the amount of soft rot that developed in the stored crop.

It must be pointed out, however, that *late* irrigation can result in wet tubers being loaded into store. Under these circumstances, the tubers will be more prone to soft rotting and will be more likely to develop gangrene (*Phoma exigua*) and skinspot (*Polyscytalum pustulans*). If irrigation is well planned, however, and stopped well before the date of haulm destruction, the risk of a wet harvest should be no more for irrigated crops than unirrigated crops.

Fertiliser Rate and Timing for Irrigated Crops

Rate of application

For many years, growers and scientists have debated whether irrigated potatoes require a different rate of fertiliser that unirrigated potatoes. High-yielding crops remove more nutrients from the soil, and it can be argued that the better yields from irrigated crops require higher rates of fertiliser. Conversely, it could be argued that a smaller amount is required because a good supply of moisture will encourage better uptake of nutrients and allow the crop to utilise fertiliser more efficiently.

There has been very little work with phosphorus and potassium in this context, but the limited evidence available supports the view that irrigation increases their availability and there is little reason to suggest that irrigated crops should require higher rates of these nutrients.

The situation with nitrogen is much more confusing, because the results from experiments have been conflicting. Staff at Levington Research Station conducted a series of experiments on commercial farms. The purpose of the experiments was to determine the optimum level of compound fertiliser at each site. The sites showed an extensive range of optima, so they were classified into three groups, viz. those receiving no irrigation, those receiving some irrigation but without efficient scheduling, and those that were properly irrigated using an irrigation scheduling service. The results are shown in Figure 5.1. The better the irrigation, the higher was the optimum amount of fertiliser. As has already been suggested, neither phosphorus nor potassium are likely to be implicated, so it was argued that the results demonstrated an effect of irrigation upon nitrogen requirement. The optimum amount of nitrogen was 180 kg/ha on the unirrigated group of trials, and 300 kg/ha on the fully irrigated crop.

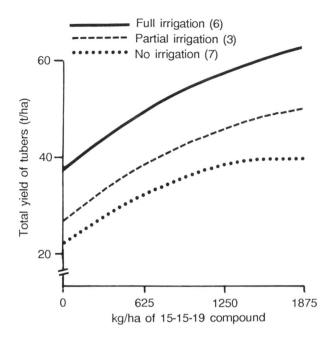

Figure 5.1   Response of potatoes to fertiliser and irrigation. Figures in brackets refer to numbers of experiments.   (Source: Levington Research Station.)

A slightly different approach was adopted by Evans of ADAS, when studying the nitrogen requirement of early potatoes. He also compared many different experiments, but as they had been conducted over a number of years, he was able to classify each result as belonging to a wet year or a dry year. He also obtained a different optimum for the two different conditions but, unfortunately, it was the reverse of the above. The optimum in dry years was 150–200 kg/ha, but in wet years it was only 100 kg/ha.

What can be learned from these contrasting results? Is it simply that irrigation has one effect on the nitrogen requirement of maincrop potatoes, but the opposite effect on that of the early crop? This is a possibility, but it is also possible that one or both groups of scientists were looking at the effects of normal site-to-site variation, or year-to-year variation, and mistakenly attributing it to a direct effect of water supply. In recent years it has become standard practice for scientists to measure the nitrogen in the soil before conducting any nitrogen experiment. This practice has developed because we now know that different fields vary enormously in the amount of residual soil nitrogen in the spring, and this variation affects results and so must be accounted for. It is quite possible that such variations, along with other site differences, are responsible for some of the effects reported in these studies.

In order to resolve this matter reliably, it is necessary to vary irrigation amounts and nitrogen within the same experiment, keeping all other factors equal. It is also necessary to conduct the experiment in a dry summer, or under an artificial rain shelter, so that there is a reasonable response to the irrigation applied. There have been several attempts to conduct such an experiment at Gleadthorpe over the years but, unfortunately, most were spoiled by the long run of wet summers that occurred in the 1960s, so that irrigation was not required in a reasonable quantity.

There are several other reports in the literature which, because they are in conflict, take us no further forward with this issue. Suffice it to say that, at present, the limited evidence suggests that the optimum level of nitrogen for irrigated maincrop potatoes following cereals on light land is 200–250 kg/ha. There is, however, a need to examine this topic again, attempting to avoid some of the pitfalls suffered by the earlier work. With this in mind, Gleadthorpe EHF and the Scottish Crop Research Institute have jointly embarked on a detailed study of this subject, and should produce some interesting results by 1990.

Timing of nitrogen application

Good irrigation scheduling should prevent over-watering, and the irrigation should not, by itself, cause a leaching problem. However, irrigated crops will be more prone to nitrogen leaching if heavy rain falls after an irrigation, particularly where an early start to irrigation is made and deficits remain low.

In order to reduce the risk of nitrogen leaching as a result of early season rainfall, a split is recommended on light soils, with one-half nitrogen in the seedbed and one-half delayed until tuber initiation. In wet springs, results with Pentland Crown at Gleadthorpe have sometimes shown a response to this splitting, and it has now become commercial practice on many light land farms. If intensive irrigation to control common scab is practised, it could be argued that there may be some advantage in withholding 50–75 kg/ha until two weeks after tuber initiation, or perhaps longer, giving a ½ : ¼ : ¼ split. This appears to be a sensible regime but, when tested at Gleadthorpe, showed no advantage over the ½ : ½ split. If growers do consider such a three-way split, it must be noted that irrigation after the final application may be essential; very late nitrogen left on a dry soil surface is not particularly effective!

It is possible to envisage other combinations of splitting nitrogen that may reduce the risk of leaching (e.g. ¼ : ½ : ¼), but further research is needed to investigate this possibility. For the time being the preferred technique involves a ½ : ½ split and, if high rainfall or faulty irrigation leads to a suspicion of leaching, obtain an assessment of the nitrogen status of the crop and supply additional nitrogen if a deficiency is indicated. On heavier soils splitting is unlikely to be worthwhile, and all the fertiliser requirements of the crop can be applied before planting.

GROWING POTATOES IN BEDS

In recent years there has been renewed interest in the concept of growing potatoes in beds rather than ridges. In early experiments beds produced many green and cracked tubers, but some of the comparisons were rather unfair as the beds were poorly made and the tubers planted at too shallow a depth.

More soil passes over the harvester when potatoes are grown in beds, and this would be a disadvantage on all but the lightest of soils. However, beds may have some advantages on such light soils, the most obvious being that the greater volume of soil in beds is likely to conserve more moisture for the crop. An earlier study at Silsoe College indicated that this was indeed the case. Following this, in 1987, researchers at Gleadthorpe EHF measured the yield advantage from using beds in both the irrigated and unirrigated situation. The results showed little difference when irrigation was applied at the full required amount, but a distinct advantage from using beds was apparent when the potatoes were unirrigated, even though 1987 was considered a wetter than average year. These results are shown in Table 5.16. The implication, of course, is that in a dry year, when water requirements are high, and growers cannot keep up with irrigation demands, growing potatoes in beds will ensure a better use of the water that is available. The results also indicate that potatoes in beds are likely to have a lower requirement for irrigation. There is a need to determine the critical deficit for potatoes grown in beds and this study will continue over the next few years.

Table 5.16 Effect of growing potatoes in beds at Gleadthorpe EHF, 1987

| | Ware yield (t/ha) | | | |
| | Pentland Squire | | Record | |
	Ridges	Beds	Ridges	Beds
Without irrigation	40.8	49.4	26.0	32.3
With irrigation	50.1	51.8	30.0	31.8

Source: Gleadthorpe EHF, ADAS.

FROST PROTECTION

The essence of growing early potatoes is to produce a high yield as early as possible. The grower will not achieve this aim if growth is checked by frost damage. There are two methods by which irrigation can be used as a defence against the worst effects of frost. The first technique involves the continuous use of sprinklers over the whole crop during the period of frost and so is not very practicable over a large area. It is sometimes used for fruit crops, however, and is described in detail in chapter 9.

The second technique is a more viable proposition for the early potato crop, and depends on the principle that wet soil will absorb and conduct heat better than dry soil. Frosty conditions in the United Kingdom are commonly associated with a weather pattern of bright, sunny days alternating with clear, frosty nights. A wet soil can absorb more heat during the daytime, and release more at night to counteract the radiational cooling. The resultant effect on temperature is quite small, the air immediately above the wet soil being about 1 °C warmer than that above dry soil. This small difference in temperature can have a marked effect on the level of damage to emerged potatoes, however.

The benefit of this technique was observed and recorded at Gleadthorpe in 1961. A field of Arran Pilot was irrigated with 25 mm of water over a period of two days, with a severe frost occurring during the intervening night. The area irrigated the day before the frost suffered from distorted leaves and blackened tips, but the area not irrigated until after the frost suffered much more severely. When the crop was harvested in June, the first area yielded 20 t/ha, but the second area only yielded 12.5 t/ha.

Further scientific work on this technique is lacking, but it is probable that a single application of water will provide some measure of protection for up to four days. An application of 8 mm every four days is recommended as a practical treatment during a period in which frosts are expected, unless the SMD is less than 8 mm, in which case application should be delayed. This is an intensive rate of application but is worthwhile if it helps avoid a serious check to a crop which is to be harvested early and is of high value.

REFERENCES AND FURTHER READING

ADAMS, M. J., CORMACK, W. and LAPWOOD, D. (1985), 'The effect of irrigation for common scab control on the incidence of powdery scab in potatoes', *Gleadthorpe Experimental Husbandry Farm Annual Review*, 35–7.

ADAMS, M. J. and LAPWOOD, D. H. (1978), 'Studies on the lenticel development, surface microflora and infection by common scab (*Streptomyces scabies*) of potato tubers growing in wet and dry soils', *Annals of Applied Biology 90*, 335–43.

EVANS, C. (1974), 'Effect of nitrogen on yield of early potatoes', *Experimental Husbandry 27*, 99–103.

FOLEY, M. F. (1985), 'A review of the effects of irrigation on the incidence of diseases', in M. K. V. Carr and P. J. C. Hamer (eds), 'Irrigating Potatoes', *UK Irrigation Association Technical Monograph 2*, 49–53.

HARRIS, P. M. (1985), 'Irrigation and the fertiliser requirements of the potato crop', in M. K. V. Carr and P. J. C. Hamer (eds), 'Irrigating Potatoes', *UK Irrigation Association Technical Monograph 2*, 25–40.

HARVEY, P. N. (1963), 'Protecting early potatoes against frost', *Agriculture 70*, 214–15.

JEFFRIES, R. A. and MACKERRON, D. K. L. (1987), 'Observations on the incidence of tuber growth cracking in relation to weather patterns', *Potato Research 30*, 613–23.

LAPWOOD, D. H., WELLINGS, L. W. and HAWKINS, J. H. (1973), 'Irrigation as a practical means to control potato common scab (*Streptomyces scabies*): final experiment and conclusions', *Plant Pathology 22*, 35–41.

LEWIS, B. G. (1970), 'Effects of water potential on the infection of potato tubers by *Streptomyces scabies* in soil', *Annals of Applied Biology 66*, 83–8.

MACKERRON, D. K. L. (1985), 'Timing of irrigation in relation to yield and quality of potatoes', in M. K. V. Carr and P. J. C. Hamer (eds), 'Irrigating Potatoes', *UK Irrigation Association Technical Monograph 2*, 54–60.

MACKERRON, D. K. L. (1986), 'Irrigation and potato quality', *Proceedings of a Potato Marketing Board Conference on 'Potato Quality'*, Coventry, UK, 83–8.

PEELER, C. H., HARVEY, P. N. and ROSSER, W. R. (1966), 'Effect of irrigation on yield and tuber diseases of maincrop potatoes', *Experimental Husbandry 14*, 30–42.

PRESTT, A. J. (1983), 'Soil management and the water use of potatoes', Ph.D. Thesis, Silsoe College, Cranfield Institute of Technology (unpublished).

TAYLOR, P. A. and FLETT, S. P. (1981), 'Effect of irrigation on powdery scab of potatoes', *Australasian Plant Pathology 10 (3)*, 55–6.

WARD, E. R. (1985), 'Irrigation in relation to disease incidence and tuber quality after storage', in M. K. V. Carr and P. J. C. Hamer (eds), *UK Irrigation Association Technical Monograph 2*, 41–8.

WELLINGS, L. W. (1973), 'The effect of irrigation on the yield and quality of maincrop potatoes', *Experimental Husbandry 24*, 54–69.

WELLINGS, L. W. and LAPWOOD, D. H. (1971), 'Control of common scab by the use of irrigation', *NAAS Quarterly Review 91*, 128–37.

WHITEAR, J. D. (1984), 'Responses of potatoes to irrigation and NPK fertiliser', *Proceedings of the North-Western European Irrigation Conference*, Billund, Denmark, 129–35.

Chapter 6

Irrigation of Sugar Beet

The rewards for irrigating sugar beet are much lower than those described for potatoes in chapter 5. First, sugar beet is only moderately sensitive to drought. Second, the crop is currently (1989) grown according to a system of quotas, and there is no guarantee of receiving a reasonable price for any yield produced over the farm's allocated A and B quota. The price for the A and B quota is fixed each year, and no increase is offered as a bonus for achieving quota in a year of shortage. Finally, there is no marked quality improvement resulting from irrigation, as is found in the case of common scab control on potatoes. The returns can, therefore, be very low. It is not surprising to find that only 12 per cent of the national sugar beet area is irrigated.

As with all other crops, the requirement for irrigation is determined by three factors:

1. The water requirement of the crop.
2. The available water capacity (AWC) of the soil.
3. Rainfall.

If the water initially stored in the soil, plus the rainfall during the growing season, is equal to or more than the requirement of the crop, no response to irrigation can be expected. Where there is a shortfall there will be a response to irrigation, and the magnitude of this response will depend on the size of the shortfall. It might be reasonable to expect, therefore, that some of the crop grown on light soils will be irrigated, particularly in dry summers, but very little elsewhere.

During 1982–5, British Sugar carried out detailed surveys of husbandry practice, including irrigation. These surveys show that there is surprisingly little relationship between soil type and the application of irrigation in a dry year (Table 6.1). As expected, a greater proportion is irrigated on the lighter soils, but one would expect this difference between soil types to be more marked.

The same survey demonstrates how much irrigation was applied in each situation (Table 6.2), which again shows very little reflection of requirements. In a dry year, crops grown on high AWC soils received

Table 6.1   The proportion of irrigated sugar beet according to soil type

Soil class	Area of sugar beet grown (ha)	Percentage irrigated in a wet year (1982 and 1985)	Percentage irrigated in a dry year (1983 and 1984)
Low AWC	30,000	15	17
Medium AWC	100,000	8	12
High AWC	70,000	2	11

Source: Dunham, Broom's Barn Experimental Station.

as much irrigation as crops grown on lighter soils. The difference in summer rainfall (see Table 6.2) between the dry years and the wet years was 100 mm, but the difference in irrigation applied was only around 25 mm. This shows that the amount of irrigation applied to the UK sugar beet crop is, in general, poorly matched to the needs of the crop.

Table 6.2   Average amounts of irrigation (mm) applied to sugar beet in UK

Soil class	Amount applied in a wet summer (mean of 1982 and 1985)	Amount applied in a dry summer (mean of 1983 and 1984)
Low AWC	69	84
Medium AWC	53	75
High AWC	54	86

Average summer rainfall (June to August) in eastern England:

1982	177 mm	1985 201 mm	Average of 1982 and 1985 189 mm
1983	68 mm	1984 110 mm	Average of 1983 and 1984 89 mm

Source: Dunham, Broom's Barn Experimental Station.

The reason for this apparent randomness in the application of irrigation to the UK sugar beet crop probably lies in the fact that, as mentioned already, the financial incentives for irrigating sugar beet are quite poor. It is very difficult to justify the capital investment required for irrigation solely on the returns from sugar beet. As a result, irrigation is usually provided primarily for another crop. Of those farms irrigating sugar beet, 80 per cent also irrigate potatoes and 35 per cent irrigate other vegetables. The amount of irrigation applied to sugar beet will depend very much on the needs of these other crops. Sugar beet will only be irrigated at such time as the potatoes and other vegetable crops are not placing demands on the system. In a dry period, farmers are too busy irrigating high value crops and do not, therefore, generally have the resources to irrigate sugar beet as required. In most cases farmers are right to have adopted this investment strategy.

As a result, the distribution of irrigated beet across soil types, and the amount of water applied to the crop, are the result of a complex interaction of factors, and not particularly associated with the water requirement of the crop.

REQUIREMENTS OF THE CROP

Work at Broom's Barn Experimental Station showed that a sugar beet crop well supplied with water uses approximately 400–480 mm of water between May and October. There can be occasional variation in this amount, depending on weather and the status of the crop, e.g. in 1973, when the latter part of the season was very hot and sunny, the crop used 500 mm. In 1974, when the crop suffered badly from virus yellows, only 300 mm of water was needed.

The demand is not evenly spread, but is much higher in July and August than in the other months. Potential evapotranspiration is normally at a maximum in June and July, but sugar beet often does not attain 50 per cent cover until early July. The indications from Broom's Barn data are that the June–August requirements are about 275 mm. Unless rainfall and the soil reserves meet this requirement, irrigation will be needed to attain potential yield.

AWC OF DIFFERENT SOIL TYPES

As explained in chapter 2, the amount of water stored in the soil and made available to the growing crop is dependent upon the type of soil. Furthermore, a soil can only yield its available water if it is well exploited by roots and so we are generally only interested in the AWC of the soil within the rooting zone, i.e. the root zone capacity.

Researchers at Broom's Barn have found that the maximum depth of water extraction by sugar beet roots was about 130 cm in mid-August and 160 cm in mid-September. By interpolation it can be assumed that about 145 cm of soil is exploited by the end of August. However, at this depth, exploitation of the soil roots was not complete, and only a portion of the AWC was extracted by the crop because of decreasing root density. The Broom's Barn experiments suggest that the depth of full rooting was about 90 cm by the end of August. If it is assumed that the crop has access to all available water to a depth of 90 cm, but only easily available water beyond that, it is possible to estimate the amount of water available to a sugar beet crop from various soil types up to the end of August (see Table 6.3). Only the silt loam can provide the 275 mm June–August requirement, even if all soils were at field capacity on 1 June. For most soils, some topping-up by rainfall or irrigation is required if the crop is to attain its potential yield.

Table 6.3   The amount of water available to a crop of sugar beet up to the end of August in various soil types

Soil type	Available water (mm)
Medium sand	105
Loamy medium sand over sand	110
Loamy fine sand over sand	125
Medium sandy loam	205
Sandy clay or sandy clay loam	195
Clay loam or silty clay loam	210
Fine sandy loam	235
Sandy silt loam	220
Silt loam	280

RESPONSE TO IRRIGATION

Sugar beet irrigation studies have been conducted at several research stations in the UK. In particular, three long series of experiments were conducted at Gleadthorpe EHF, Broom's Barn and Woburn Experimental Farm in Bedfordshire. The yield response from each experiment is shown in Figure 6.1, along with the associated June to August rainfall (mm). As expected, the relationship between yield response to irrigation and summer rainfall is not perfect, owing to soil variation between sites, variation in transpiration demand between years and differences in the distribution pattern of the rainfall in spring and summer. However, despite all of these potential causes of variation, a broad relationship is still clear. If the summer rainfall exceeds 140 mm, the yield response is likely to be of the order of 1.0 t/ha sugar or less, and higher responses are generally associated with less rainfall. Yield responses greater than 1 t/ha sugar can only be reliably expected when the summer rainfall is less than 100 mm. The average June–August rainfall throughout the sugar beet growing areas of the

UK varies between 160 and 200 mm. In the majority of years it is therefore expected that the yield response to irrigation will be less than 1 t/ha sugar.

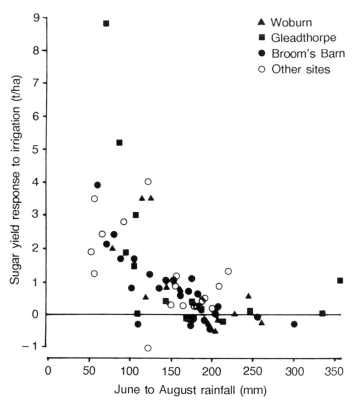

Figure 6.1   Effect of summer rainfall on the irrigation response of sugar beet.   (Source: Dunham, Broom's Barn Experimental Station.)

Of the three sites with a long series of experiments, the lightest soil is found at Gleadthorpe EHF, where the soil is a loamy medium sand over sand, and where data have been accumulated over twenty years. The yield responses are given in Table 6.4 and summarised in Table 6.5.

The very high responses recorded in 1955 and 1959 are interesting. They are almost certainly due to the exceptionally low June–August rainfall in those two years, only 88 and 73 mm respectively, compared with the average of 176 mm recorded over the whole twenty-year period.

Table 6.4 Yield response from irrigation of sugar beet at Gleadthorpe EHF

Yield	Irrigation applied (mm)	Yield response (t/ha clean beet)	(t/ha sugar)
1955	175	31.6	5.6
1956	0	0	0
1957	25	−0.5	−0.3
1958	78	2.7	1.0
1959	250	45.9	8.7
1962	74	−1.7	0
1963	89	0.7	−0.1
1964	131	6.0	0.4
1965	50	1.8	0.2
1966	50	1.5	0.1
1970	100	12.8	1.8
1979	124	13.2	2.8
1980	30	3.3	0.6
1981	94	8.5	2.8
1982	80	6.0	1.0
1984	105	16.4	3.2
1985	98	4.2	0.1
1986	91	4.0	0.7
1987	46	−1.8	−0.3
1988	80	2.1	0.5

Source: Gleadthorpe EHF, ADAS.

Table 6.5 Irrigation of sugar beet at Gleadthorpe EHF, 1958–88

June–August rainfall (mm)	Clean beet yield increase (t/ha)	Sugar yield increase (t/ha)	Water applied (mm)
Over 140 (13 years)	2.31	0.31	65
100–140 (4 years)	9.10	2.19	99
Under 100 (3 years)	30.10	5.37	175
20-year mean	7.84	1.44	89

Source: Gleadthorpe EHF, ADAS.

The data show the level of yield response that has been associated with three levels of summer rainfall, viz. under 100 mm, 100–140 mm and over 140 mm. In order to obtain a more balanced view of the frequency with which summer rainfall fits into the various categories, it is useful to look at long-term weather data at Gleadthorpe. In the last 36 years, 21 years have shown a summer rainfall greater than 140 mm, 7 years under 100 mm and 8 years in the middle range. Given this frequency of obtaining yield responses of the order of those shown in Table 6.5, the average 36-year long-term irrigation response can be estimated at about 9 t/ha clean beet and about 1.7 t/ha sugar. Not many farmers could achieve this on a commercial scale because very large amounts of water were applied in the driest years, e.g. in 1959 250 mm were applied to obtain the yield response of 8.7 t/ha sugar. Farmers with less water available would naturally expect a lower response in that year, and hence a lower average response over a run of years.

It has already been stated that most growers are too busy irrigating potatoes and other vegetables in dry years to give sugar beet any priority. Any expectation of yield response from sugar beet must take into account the maximum amount of water available for irrigation of the beet crop in a dry year. The principle is shown in Table 6.6.

Table 6.6   Estimates of sugar beet irrigation response at various levels of irrigation capacity at Gleadthorpe EHF

Maximum amount of irrigation available in a dry year (mm)	Average amount used (mm)	Estimated average response (20 years)	
		(t/ha clean beet)	(t/ha sugar)
25	24	1.5	0.3
50	45	3.0	0.6
75	62	4.3	0.8
100	74	5.5	1.0
125	79	6.2	1.1
150	82	6.7	1.2
175	87	7.1	1.3
200	86	7.4	1.4
225	87	7.6	1.4
250	89	7.84	1.44

The responses to irrigation at Broom's Barn and Woburn have been considerably lower than those obtained at Gleadthorpe. A comparison of average yield response at the various sites is shown in Table 6.7 along with a measure of the efficiency of irrigation as given by the yield response per 25 mm applied.

Table 6.7 Effect of irrigation on sugar yield at various UK sites

Site	Number of years covered by experiments	Average sugar yield increase	
		t/ha	t/ha per 25 mm irrigation
Gleadthorpe	20	1.44	0.40
Broom's Barn	23	0.79	0.17
Woburn	12	0.89	0.24
Other sites	18	1.17	0.24

Source: Dunham, Broom's Barn Experimental Station.

Researchers at Broom's Barn have measured the water use efficiency of sugar beet and shown that the crop produces 1 t/ha sugar for each 25 mm of *water used*. However, as shown in Table 6.7, the average efficiency of *irrigation* at Gleadthorpe has been slightly less than half of that, and at other sites it has been much less. There appears to be a large discrepancy between irrigation applied and water actually used. In fact, the inference is that only 40 per cent of irrigation water at Gleadthorpe is actually taken up by the crop, and only 17–24 per cent elsewhere. These average figures for irrigation efficiency cover a large amount of variation. Table 6.8 shows the irrigation efficiency obtained at Gleadthorpe with various summer rainfall amounts. The results show that irrigation efficiency is much higher in drier years, i.e. the large amounts of water used in dry years are used more efficiently than the smaller amounts used in wet years. There are various possible reasons for this.

Table 6.8 Irrigation efficiency at Gleadthorpe EHF

June–August rainfall (mm)	Average sugar yield increase (t/ha per 25 mm irrigation)
Over 140 (13 years)	0.12
100–140 (4 years)	0.55
Under 100 (3 years)	0.77

Source: Gleadthorpe EHF, ADAS.

1. In wet years, the likelihood of rain falling soon after irrigation has been applied is naturally higher, and this will reduce the need for the irrigation, and hence its effective use.

2. As explained in chapter 2, wet years are associated with lower transpiration demand for water. The degree of moisture stress experienced by the crop depends on the balance between transpiration demand and rate of supply from the soil. Rate of supply from the soil is influenced by the SMD. When the rate of transpiration demand is low, the rate of supply need only be low and crops can withstand higher SMDs than are possible in a hot summer without any detrimental effect on yield. As the experiments have used the same trigger deficits in both wet and dry years, it follows that the plots received more water than was strictly necessary in the wetter years. This will contribute to a lower efficiency of irrigation in such years.

In summary, the yield response on light soils in dry years is around 15 t/ha clean beet or 3 t/ha sugar, but on average the response over a number of years is around 8 t/ha clean beet or 1.5 t/ha sugar, or somewhat less if irrigation is limited. This average depends very much on the high responses in dry years. In most summers, when over 140 mm rain falls between June and August, the responses are much lower. On heavier soils, such as Broom's Barn where the AWC is greater, a smaller response can be expected. Where information is required for investment analysis on a particular farm, much better data can be obtained by analysing the soil and local weather pattern as described in chapter 3.

Of course, with the present quota system operating for sugar beet, there is no real incentive to increase yields above quota, and any financial assessment of beet irrigation must take that factor into account. The main advantage of irrigation is that it can be used to avoid major shortfalls in yield, allowing a more reliable calculation of the area required to meet the A and B quota fully, and thereby optimise land use.

IRRIGATION SCHEDULING FOR SUGAR BEET

In 1984 and 1985, British Sugar conducted surveys of irrigated beet crops and the method used to schedule the irrigation. In all, 129 fields of beet were involved in the survey, and the results are shown in Table 6.9. The results showed that some form of water balance calculation (manual or computer-based) was the commonest technique. It is also interesting to note that while only 39 growers described their method as being 'very satisfactory', most of these were using a water balance calculation. The category 'other methods' in Table 6.9 refers to situations where scheduling was not really applicable, e.g. irrigating to assist emergence through a surface crust.

Table 6.9   Survey of irrigation scheduling methods used for sugar beet, 1984–5

Method	Number of surveyed fields where the method was used	Users' appraisal of the method (number of fields in each category)			
		'Very satisfactory'	'Quite useful'	'Little help'	No comment
Water balance	75	28	44	2	1
Degree of wilting	43	6	22	6	9
Feel of soil	23	3	15	0	5
Tensiometers	3	0	2	1	0
Other methods	11	2	6	0	3

Source: Dunham, Messem and Turner, Broom's Barn Experimental Station.

As described in earlier chapters, the water balance method entails calculating the SMD every day (usually by a computer nowadays) and irrigating before the critical SMD is reached. What is the critical SMD for sugar beet? This depends very much on soil type. We can allow a higher SMD to develop in a loamy soil than we can in a sand, before yield is affected. The rooting depth of the crop is also significant. A crop with shallow roots early in the season is much more sensitive to drought, and has a lower critical SMD, than the same crop two months later when it has a deeper rooting system. This is a particularly important feature of sugar beet, because the rooting system develops so much during the growing season. In June, the roots are shallow and sparse, but by September, the roots are well developed and deeper than most of our other crops. All these factors have to be taken into account in the estimation of critical SMDs. Examples of critical deficits for sugar beet are shown in Table 6.10.

Table 6.10   Critical soil moisture deficits (mm) for sugar beet

	Coarse sands	Loamy sands	Sandy loams	Clay loams
June	25	30	35	50
July	35	40	50	100
August	50	60	75	125
September	65	75	125	150

Source: Broom's Barn Experimental Station.

A good irrigation schedule should be designed to keep the SMD below these critical levels throughout the season. As an example, on a loamy sand, a typical irrigation plan in June would be to apply 25 mm whenever the SMD approaches 30 mm. In September, however, the SMD can be allowed to approach 75 mm before irrigation need be applied. These critical deficits are good guidelines for the soils shown, but many soils are combinations of different textures in the topsoil and subsoil. For a full description of how to calculate critical deficits for different crops and different soil types, see chapter 2.

Linking critical deficits to growth stage might be better than keeping strictly to the month. This is suggested by the fact that the progress of development of the crop in the early part of the season shows significant variation from year to year. For example, the season could be divided up as follows:

Period 1   Up to 50 per cent crop cover.
Period 2   50 per cent crop cover to end of July.
Period 3   August.
Period 4   September.

This would improve matters as it would eliminate the problem of applying unrealistic critical deficits to unusually advanced or unusually late crops. There is probably no need to link the recommendations in August and September with growth stages, because crops have usually evened out by early August.

The critical deficits on sands and loamy sands in June are low, and are likely to be exceeded in the majority of years unless irrigation is applied. But the majority of growers with irrigation also have potatoes and/or field vegetables to irrigate, so priorities have to be considered. Even though beet can be quite sensitive to drought at this time, it is inevitable that potatoes and field vegetables should take priority because of their high value, and the higher returns obtained from irrigating them. For a fuller discussion of priorities, see chapter 12.

EFFECT OF WATER ON SUGAR PERCENTAGE

In general, the experiments at Gleadthorpe have shown very little effect of irrigation on sugar percentage.

High sugar percentages are generally associated with summers of low rainfall, as shown in Table 6.11. However, in such years, the irrigated crops also show high sugar levels, so the increased sugar

Table 6.11 Effect of irrigation on sugar percentage at
Gleadthorpe EHF, 1958–88

June–August rainfall (mm)	Percentage sugar		Irrigation applied (mm)
	Unirrigated	Irrigated	
Over 140 (13 years)	17.1	17.0	65
100–140 (4 years)	17.2	17.5	99
Under 100 (3 years)	18.2	17.9	175
20-year mean	17.3	17.2	89

Source: Gleadthorpe EHF, ADAS.

cannot be a direct result of the low water availability in such years. More likely, the high sugar levels result from the high levels of sunshine that are associated with low rainfall years. Irrigation is therefore unlikely to result in a marked decrease in sugar percentage.

Broom's Barn has also shown that irrigation in June–August had little effect on sugar percentage, but irrigation in September decreased sugar percentage in their experiments by 0.5 percentage units. In 1985, results at Gleadthorpe supported this view, as shown in Table 6.12.

Table 6.12 Effect of irrigation at different times on sugar percentage at
Gleadthorpe EHF, 1985

Irrigation	Percentage sugar	Amount water applied (mm)
Nil	18.9	—
May	18.8	11
May–June	18.8	30
May–July	18.6	91
May–July + September	18.2	149

Source: Gleadthorpe EHF, ADAS.

This is not necessarily a consistent effect, however, as results at Gleadthorpe in 1986 and 1987 showed no decrease.

WHEN TO START AND STOP IRRIGATING

When to start

The critical deficits shown in Table 6.10 take into account the fact that the crop has a shallow and sparse root system in June. In the month of May, however, the root system is even more restricted. In warm dry springs, the soil can dry out in the immediate vicinity of the roots. Should we irrigate the crop in a dry May? Is there a critical deficit for May that should be included in Table 6.10? A few experiments have investigated this, but not a single result shows irrigation in May to be of benefit to the crop. The results of an experiment at Gleadthorpe in 1987 can be used as an illustration of this point. May 1987 was a very dry month, with only 64 per cent of the long-term average rainfall for the month, i.e. 34 mm. The SMD was 20 mm on 1 May, and increased throughout the month on unirrigated plots. Even under these conditions, applying 18 mm irrigation to some plots on 1 May had no effect on the final yield of the crop. Various explanations are possible for this:

1. The low crop covers and the low transpiration rates, even in so dry a May, are such that the crop could meet the demands.
2. The small root system is continually expanding into previously unexploited soil moisture.
3. It is possible that the unirrigated plots did suffer stress, but their resumed growth in the very wet June subsequently compensated for this stress. There is much debate over this last possibility, as some scientists claim that sugar beet can compensate for early season stress, but others are equally adamant that the crop does not have the potential to do this.

Whatever the explanation, all we can state with certainty is that the crop showed no yield response to irrigation in a dry May. It is probable that there are some seasons when the crop would show a response, but they are probably rare occasions, and the precise conditions are not yet known. For the present, I would only irrigate the crop if the SMD were above 20 mm *and* there was a forecast of dry, hot and sunny conditions for the next seven to ten days. This assumes that the irrigation can be applied via a boom irrigator or equivalent. The risk of damaging small seedlings with a raingun is high and could cancel the beneficial effect of increased water availability.

When to stop

Irrigation at the end of the season has been investigated in other experiments. Experiments at Broom's Barn showed that irrigation in September increased fresh root weight, but was accompanied by a proportionate decrease in sugar concentration. Consequently, there was no overall loss or gain of sugar yield as a result of water given in September.

An experiment at Gleadthorpe in 1985 gave different results (see Table 6.13). The season was fairly wet until the end of August, and was followed by a very dry autumn. The rainfall in September was only 24 per cent of the long-term average. In such a wet season, there was no response from irrigating the crop in May, June or July. The SMD at the beginning of September was 60 mm, and by the end of the month had exceeded 100 mm where irrigation was not applied. September irrigation in these circumstances produced a yield increase of around 5 t/ha of clean beet. Although this was partially offset by a decrease in sugar concentration (see Table 6.13), there was still a resulting increase in sugar yield of around 0.5 t/ha. It is therefore concluded that irrigation of sugar beet in September is a worthwhile practice in seasons when the deficits are particularly high. This advice is taken into account in the critical deficits shown in Table 6.10, and also in the detailed account of critical deficit estimation given in chapter 2.

Table 6.13 Effect of irrigating in September at Gleadthorpe EHF, 1985

Irrigation	Yield of clean beet (t/ha)	Percentage sugar	Sugar yield (t/ha)	Amount of water applied (mm)
Nil	41.0	18.9	7.76	—
May	39.2	18.8	7.39	11
May–June	39.6	18.8	7.44	30
May–July	42.4	18.6	7.89	91
May–July + September	46.0	18.2	8.37	149

Source: Gleadthorpe EHF, ADAS.

If irrigation in September is worthwhile, are there some circumstances in which irrigating in October would be useful? This was tested at Gleadthorpe in 1986, and the results are presented in Table 6.14. Over 70 mm rainfall occurred within 48 hours in late August. Following this, September and early October were particularly dry, with only 7 mm rainfall in 6 weeks. In this situation, extending irrigation into

October was compared with stopping earlier. There was no yield increase obtained from the extra application in October (treatment 3 compared with treatment 2).

There is another aspect to this late-season irrigation that requires discussion. In particularly dry years, growers are so busy irrigating potatoes that the sugar beet often gets left with little or no irrigation and SMDs can reach very high levels. In such years, farmers question whether some late-season irrigation, after the potatoes are finished with, is beneficial to their sugar beet. In order to investigate this possibility, the 1986 experiment included plots that had been left unirrigated all summer, but were irrigated intensively in September, early October or late October (treatments 4, 5 and 6). The potential SMD on unirrigated plots exceeded 150 mm in October. Under these circumstances late irrigation resulted in 4—7 t/ha extra clean beet. Although this was once again partially offset by a decrease in sugar percentage, there was still a resulting increase in sugar yield of 0.5—1.0 t/ha.

Table 6.14   Effect of irrigating in October at Gleadthorpe EHF, 1986

Irrigation	Clean beet yield (t/ha)	Sugar yield (t/ha)	Irrigation applied (mm)
1. Nil	43.3	7.95	—
2. June—September	46.4	8.58	85
3. June—October	47.0	8.67	109
4. September only	47.1	8.45	73
5. 1—15 October only	50.0	8.97	95
6. 16—31 October only	47.6	8.71	82

Source: Gleadthorpe EHF, ADAS.

The result shows that irrigation of sugar beet in October can pay dividends on very light soils when the crop has received little previous irrigation and soil moisture deficits are high. Thus it does appear that growers can, by irrigating in the autumn, recover *to some extent* the yield lost as a result of withholding water from the beet crop during the main irrigation season when potato irrigation takes priority. In this particular experiment, the crop irrigated only in October appeared to recover lost yield in full, and this treatment gave as good a yield as the crops irrigated throughout the season! I would be surprised to see quite such a favourable response again in another year. Work on this is continuing.

The state of the crop will also surely affect the response. If an early drought had been so severe as to reduce the leaf canopy markedly, the capacity of the crop to respond to late irrigation must be limited.

It must be pointed out that many abstraction licences only allow for irrigation to be applied up until the end of September. If growers intend to irrigate in October, they should first consult with their water authority and seek permission to do so.

EFFECT OF IRRIGATION ON NITROGEN REQUIREMENT AND TIMING

Experiments on nitrogen response, by Broom's Barn Experimental Station and Gleadthorpe EHF, have generally shown, on a range of light soils, that the irrigated crop has a similar nitrogen requirement to the unirrigated crop, with 125 kg/ha normally being adequate for maximum yield (see Figure 6.2). However, nitrogen experiments on irrigated sandland, i.e. at the lightest end of the range, have sometimes shown a higher optimum requirement. This has led to some controversy regarding the amount of nitrogen required when irrigation is applied.

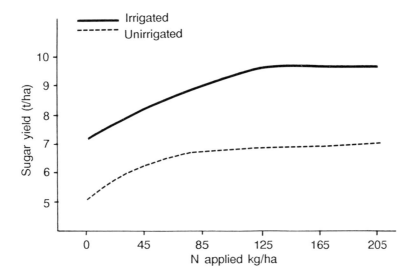

Figure 6.2   Nitrogen response of sugar beet with and without irrigation.   (Source: Gleadthorpe EHF, ADAS.)

To obtain better information on the general subject of the nitrogen requirements of sugar beet, a joint study involving staff from British Sugar, Broom's Barn and Rothamsted has recently been initiated. The study involves monitoring nitrogen requirement carefully at twenty-three sites. Preliminary results indicate that the optimum nitrogen dressing ranges from 150 kg/ha on some sandy soils to 0 kg/ha elsewhere. The higher optimum on the sandy sites does not appear to be related to irrigation, however, but to the low levels of residual nitrogen found in the soil at such sites, i.e. it is a result of the soil type, not of irrigation. This is very interesting as it allows the possibility of determining the optimum level of nitrogen in advance by analysing soil samples taken in early May. The study is continuing at present.

There are, of course, some conditions under which irrigation is likely to increase the required rate of applied nitrogen. In years of excessive spring rainfall, some nitrogen is inevitably leached beyond rooting depth. While irrigation should not, by itself, cause leaching if it is accurately scheduled, it does increase the risk. If heavy rain falls after irrigation has been applied, there will be more drainage than found in an unirrigated crop, and hence more leaching. If this leaching has been severe, it may be justifiable to apply a small extra dose of nitrogen. It is not justified to do this as an insurance, however, and should only be applied after seeking advice.

Effect of irrigation on growth of roots

It has been argued by some that roots of all crops grow in active response to a need for water, and so a crop that is well supplied with water in the early stages will develop a shallow and sparse root system. Following this line of argument, an early irrigated crop might be less able to extract its nutrient requirement from the soil, and less able to extract moisture from depth if irrigation were to be interrupted. The evidence for this is not at all conclusive and, in fact, sometimes suggests the opposite.

Studies of sugar beet roots have produced some interesting findings. The first reports were those of Weaver in the USA, during the 1920s. His studies (Figure 6.3) showed that the crop produced a deeper rooting system when the soil was fully irrigated than when water was restricted. Much more recently, sugar beet rooting has been studied at Broom's Barn. Staff there have found that an early drought (June and July) had severe effects on the rooting system, and led to decreased rooting in both upper and lower layers. Root growth at depth was increased by irrigation. However, the same studies also revealed that late drought (August and September) did appear to increase rooting at depth, although the magnitude of the effect was small (see Figure 6.4).

The effect of the timing of the drought could be responsible in part for what appears to be a conflict of reports in the literature. It must also be remembered that over-watering can create anaerobic conditions which inhibit root growth. This may lead different workers to different conclusions. Some workers may have recorded the beneficial effects of adequate watering upon root growth in a dry season. Yet others may have recorded the detrimental effects associated with irrigating excessively or, indeed, with heavy rainfall following sufficient irrigation. It is easy to see how a conflict of opinion could arise!

Experimental evidence from other crops shows that irrigated plants can have better root systems. Experiments with wheat at Rothamsted and apple trees at East Malling have shown that crops with adequate water produce better rooting systems than droughted crops (see Table 9.9).

An experiment with potatoes at Gleadthorpe EHF in 1988 has provided further information. As explained in chapter 5, it is common practice with potatoes to irrigate intensively during the six weeks immediately following tuber initiation, applying 12 mm water whenever the SMD reaches 15 mm. This intensity of irrigation is certainly far greater than that required by the crop for normal growth, but is applied in order to control the disease common scab. At the end of the six-week period, the intensity of irrigation is reduced and matched to the normal demands of the crop. In 1988, a study of rooting patterns showed that intensively irrigated plots produced a shallow rooting system early in the season relative to unirrigated plots. However, when the irrigation intensity was reduced, six weeks after tuber initiation, roots then extended quite rapidly and eventually overtook those of the unirrigated plots.

In general it is likely that there is an optimum level of soil moisture for prolific root growth, and that this level is best obtained by irrigating according to requirement with an accurate scheduling technique. Under-watering in the early stages is likely to inhibit root growth, as will over-watering. In summary, properly scheduled irrigation is unlikely to decrease the availability of nutrients through leaching or restricted root growth. In fact, quite the opposite is true: irrigation is often likely to increase the availability of nutrients through increased rooting and by improving the diffusion of nutrients through the soil to the roots.

Timing of nitrogen application

Nitrogen applied in the seedbed can have an adverse effect on newly germinating seeds, which can in turn result in poor emergence of the

Figure 6.3   *Root systems of sugar beet with and without irrigation. (Source: Adapted from Weaver, 1926.)*

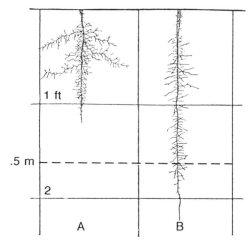

About 2 months old: A, dry land (practically no water available in the second foot); B, irrigated soil.

crop and a lower plant population. Deflecting the nitrogen away from the rows, by using band application, should avoid this problem but is not a commonly used method. Even with this technique, there is still the possibility of heavy rains leaching some of the nitrogen. Both risks can be reduced by applying 40 kg/ha nitrogen overall at (or just after) drilling and the remainder after full emergence at the 2–4 true-leaf stage of growth. This system has been shown to be very success-ful with irrigated crops, and is widely used. The results of nitrogen timing trials at Gleadthorpe are presented in Table 6.15. In recent years there is evidence to suggest that application at the 2-leaf stage is slightly preferable to the 4-leaf stage.

Table 6.15   **Effect of rate and timing of nitrogen on sugar yield of irrigated sugar beet at Gleadthorpe EHF, 1979–81**

Timing		Sugar yield (t/ha)		
	N kg/ha	125	165	205
Seedbed		9.2	9.9	9.5
Post-drilling overall		9.8	9.8	9.6
Post-drilling deflected away from rows		9.5	10.1	9.4
40 kg/ha in seedbed, remainder post-emergence (4 true-leaf stage)		9.8	9.9	9.9

Source: Gleadthorpe EHF, ADAS.

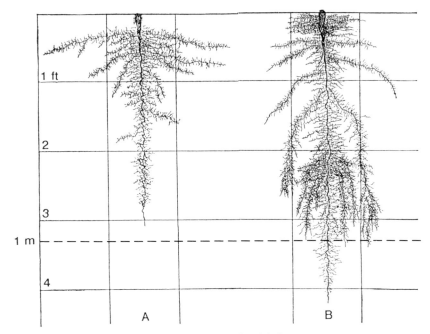

About 3 months old: A, dry land with low water content of subsoil; B, fully irrigated soil.

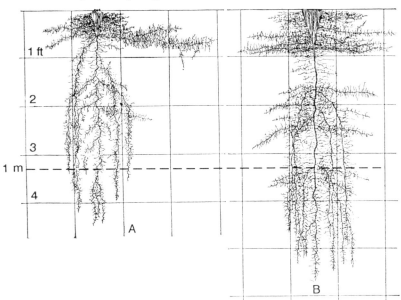

On Sept. 12: A, dry land; B, fully irrigated soil.

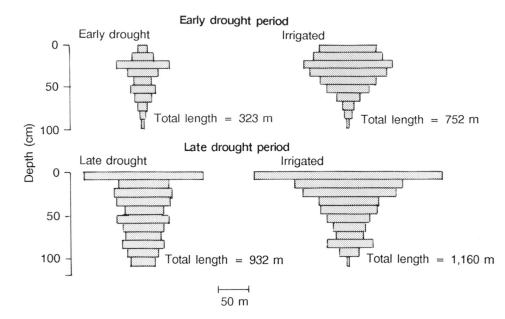

Figure 6.4   Effect of early and late drought on sugar beet roots.
(Source: Broom's Barn Experimental Station.)

Irrigation and amino nitrogen

The application of too much nitrogen fertiliser will lead to higher levels
of amino nitrogen in the root. Amino nitrogen is an impurity that
reduces the efficiency of sugar extraction at the factory. If levels exceed
150 mg amino nitrogen per 100 g sugar on mineral soil or 250 mg
on organic soils, the amount of nitrogen fertiliser used should be
reviewed.

Levels of amino nitrogen can also be affected by irrigation. Experiments at Gleadthorpe have investigated this. Results show that amino
nitrogen levels were reduced when the crop was irrigated (see Table
6.16).

OTHER EFFECTS OF IRRIGATION

Irrigation is likely to have an effect on various pests and diseases, among
which are virus yellows and *Rhizomania*.

**Table 6.16   Effect of nitrogen and irrigation on amino nitrogen
(mg per 100 g sugar)**

		Nitrogen applied (kg/ha)				
		45	85	125	165	205
1979	Unirrigated	115	106	126	163	183
	Irrigated	84	90	114	124	154
1981	Unirrigated	79	95	109	138	167
	Irrigated	67	86	104	124	148

Source: Gleadthorpe EHF, ADAS.

Virus yellows

Virus yellows is the most important disease of sugar beet. It is a virus disease caused by either of two viruses — beet yellow virus (BYV) or beet mild yellowing virus (BMYV).

The disease is spread by aphids. Some of the worst epidemics of virus yellows have occurred in dry summers. In general, the disease is less prevalent in wet summers and this is believed to result from the fact that heavy rainfall either drowns the aphid vector, or at least washes it off the sugar beet leaves. Irrigation is often applied in very heavy doses (commonly 25 mm per application) which are extremely effective in washing aphids away, as can easily be seen by walking through a crop before and after treatment. It is therefore expected that irrigation should offer some protection against late infection in a dry year, but unfortunately much of the infection occurs before the start of the irrigation season.

Rhizomania

Rhizomania is a disease caused by the Beet Necrotic Yellow Vein Virus (BNYVV). It is carried by a common soil-borne fungus, *Polymyxa betae*, can spread very rapidly and may persist in the soil for many years.

The first outbreak of *Rhizomania* in the UK was reported in 1987, followed by two subsequent reports in 1989, and it has become strongly established in some areas of Europe. Symptoms are variable, depending on the time of infection, but often include taller plants with narrow leaves, long stalks and a severely restricted tap root.

Sometimes a 'beard' of fibrous roots appear. Yield is severely restricted, and beet growing is usually uneconomic in areas where infection is at a high level.

Irrigation is probably responsible for increasing the speed with which *Rhizomania* has spread in some countries. The fungus prefers wet soil conditions, which enable it to grow and spread quickly.

Irrigation run-off from a contaminated field is likely to contain the fungus and the virus. If this were to enter a water supply which is used to irrigate other sugar beet fields, spread is quite possible.

Irrigation to aid establishment

Germinating seedlings may benefit from *small* amounts of irrigation to bring them through dry soil caps, if these develop. This is a technique that is only recommended in extreme cases, however, and should be considered as a last resort. There is a risk that any emerged seedlings will be damaged. If the weather remains dry, the crust may develop again, and it could be harder than before. The amount applied should be sufficient to wet the surface layer, but saturating the soil around the seed should be avoided. In any circumstances, 12 mm water should be sufficient, and possibly excessive, but applying smaller amounts with a raingun is not practicable. If the weather remains dry, it may be necessary to repeat the application.

Irrigation to aid harvest

In a dry autumn, some soils may become extremely hard. A small amount of irrigation may serve to soften the ground but over-watering should be avoided. Transpiration rates are low in the autumn, and excessively wet soil could take a long time to dry out.

REFERENCES AND FURTHER READING

ARCHER, J. (1988), *Crop Nutrition and Fertiliser Use*, 2nd edition (Farming Press), pp. 265.

BAILEY, R. J. (1986), 'Yield response of sugar beet to irrigation at Gleadthorpe EHF 1958—85', *Irrigation News 10*, 45—7.

BRAY, W. E. and THOMPSON, K. J. (1985), 'Sugar beet: a grower's guide', 3rd edition, Sugar Beet Research and Education Committee, 59—63.

BROWN, K. F., MESSEM, A. B., DUNHAM, R. J. and BISCOE, P. V. (1987), 'Effect of drought on growth and water use of sugar beet', *Journal of Agricultural Science, Cambridge 109*, 421—35.

DRAYCOTT, A. P. and MESSEM A. B. (1977), 'Response by sugar beet to irrigation 1965—75', *Journal of Agricultural Science, Cambridge 89*, 481—93.

DUNHAM, R. J. (1987), 'Irrigating sugar beet in the United Kingdom', *Proceedings of the 2nd North-Western European Irrigation Conference*, Silsoe, UK.

DUNHAM, R. J. (1988), 'Irrigation of sugar beet: the main effects', *British Sugar Beet Review 56 (3)*, 34–7.

HARVEY, P. N. and WELLINGS, L. W. (1970), 'Irrigation of sugar beet on light sand soil', *Experimental Husbandry 19*, 1–12.

HOLMES, M. R. J. and WHITEAR, J. D. (1976), 'Nitrogen requirement of sugar beet in relation to irrigation', *Journal of Agricultural Science, Cambridge 87*, 559–66.

JAGGARD, K. W., FARROW, B. and HOLLOWELL, W. (1989), *Sugar Beet*: A Grower's Guide, 4th edition, Sugar Beet Research and Education Committee, 70–4.

LAST, P. J., DRAYCOTT, A. P., MESSEM, A. B. and WEBB, D. J. (1983), 'Effects of nitrogen fertiliser and irrigation on sugar beet at Broom's Barn 1973–78', *Journal of Agricultural Science, Cambridge 101*, 185–205.

PENMAN, H. L. (1952), 'Experiments on irrigation of sugar beet', *Journal of Agricultural Science, Cambridge 42*, 286–92.

WEAVER, J. E. (1926), *Root Development of Field Crops* (McGraw-Hill).

Chapter 7

Irrigation of Combinable Crops

CEREALS

The previous two chapters have highlighted the fact that irrigation of sugar beet, although worthwhile in some situations, does not produce the high financial returns that are commonly associated with irrigation of potatoes. Irrigation of cereals is even less likely to be financially worthwhile, particularly with the expectation of lower prices for cereals in the early 1990s. Nevertheless, the MAFF survey of 1984 showed that in a dry year UK farmers irrigated 25,000 ha of cereals, which represented 17½ per cent of the total irrigated area within the UK. Clearly, many farmers believe that cereal irrigation is worthwhile. In this section, the economics of irrigating cereals will be closely examined.

Yield response

Deep rooting crops, such as cereals, when grown on a soil with a high available water capacity, are unlikely to benefit from irrigation except in extremely dry years. This view is reinforced by work on the silt loam soil at Letcombe Laboratory, where it was shown that an SMD of 150 mm, expected in only about 15 years in 100, had a small and non-significant effect on the yield of winter wheat. There were measured decreases in the number of ears, and the number of grains per ear, as a result of the drought, but these were offset by an increase in grain size.

Similarly, winter wheat on the silty clay loam at Rothamsted in 1972 and 1974 did not respond to irrigation, and the workers suggested that yield was unlikely to be decreased unless the SMD exceeded 150 mm on that soil. This conclusion must be modified, however, as a result of an experiment in 1985 on this soil, when drought created artificially by a mobile crop shelter produced large effects on winter wheat, with transpiration being reduced once the SMD exceeded 100 mm.

Experiments on the loamy medium sand at Gleadthorpe EHF have sometimes shown large yield responses from irrigation (see Colour plates 17 and 18). Yield responses over the period 1981—8 at Glead-thorpe are shown in Table 7.1. In two experiments (winter wheat in 1981, winter barley in 1983), irrigation was not applied due to the exceptionally wet weather prevailing. The response in these two experiments has been taken as zero.

Table 7.1 Yield response from irrigation of cereals at Gleadthorpe EHF

Crop	Year	Irrigation applied (mm)	Yield response (t/ha)
Winter wheat	1981	0	0.00
	1982	127	1.00
	1983	20	0.53
	1984	107	2.27
	1985	25	−0.26
	1986	51	−0.18
	1987	74	−0.26
	1988	67	−0.07
Winter barley	1982	70	0.46
	1983	0	0.00
	1984	75	1.26
	1985	56	0.18
	1986	34	0.27
	1987	94	0.37
	1988	88	−0.02
Spring barley	1985	40	0.44
	1986	70	0.47
	1987	52	−0.07
	1988	62	0.16

Source: Gleadthorpe EHF, ADAS.

Before examining these responses in closer detail, it is necessary at this juncture to consider whether winter wheat, winter barley and spring barley show similar responses to irrigation or whether they are affected differently.

The data in Table 7.1 show no consistently larger response from any one cereal. Indeed, it would be surprising if one of these cereal species consistently showed a greater irrigation response than others. Various factors are involved in determining the response to irrigation, including the soil on which the experiment was conducted, the depth of rooting at the time of moisture stress, stage of growth, effect of irrigation upon subsequent lodging, etc. The variable pattern of the UK climate, with moisture stress occurring at different times, will inevitably result in a different order of response between cereal species in different years.

It is, nevertheless, possible to draw some conclusions about different cereal species. It is widely supposed that rye is more resistant to drought than other species and, because of this, it is commonly grown on light soils in East Anglia. Several experiments at Gleadthorpe have shown that unirrigated rye will generally outyield unirrigated wheat. Also three experiments in the 1960s at Gleadthorpe showed that spring oats were much more responsive to irrigation than spring barley.

The explanation frequently given for such differences is that cereal species differ in rooting depth, with rye being particularly deep rooting, but there is little experimental evidence to support this view. Indeed, a study of wheat and rye rooting patterns at Gleadthorpe in 1988 showed very little difference but a slight tendency for wheat to root more deeply.

Effective rooting of cereals is, however, often restricted by the disease Take-All. Cereal species differ in their susceptibility to this disease, as shown in Table 7.2, with rye generally having lower levels of infection than wheat or barley. This may contribute to the greater degree of drought tolerance shown by rye. This explanation does not account for oats, however, which, although immune to the strain of Take-All found in eastern parts of the country, nevertheless exhibit a low tolerance of drought.

Table 7.2 Average index of Take-All infection (1986–8)

Cereal	Date of assessment	
	June	July
Wheat cv Avalon	106	181
Barley cv Panda	84	127
Triticale cv Lasko	76	106
Triticale cv Status	79	95
Rye cv Domino	38	41

Source: Rothamsted Experimental Station (Prew, personal communication).

Another explanation is offered by differences in the pattern of growth. Table 7.3 shows the stage of growth attained by the various species at Gleadthorpe on 12 May 1988. Rye was the most advanced, followed in turn by winter barley, triticale, winter wheat and oats. This difference in maturity continued throughout the season and the order of flowering was similar. While the UK climate is variable, there is a general tendency for the highest SMD to be recorded towards the end of the summer. It follows that an early maturing crop such as rye is likely to avoid the worst effects of drought in many years by passing through the particularly sensitive stage of flowering before moisture stress becomes limiting. On the other hand, a later maturing crop such as oats will be subjected to moisture stress more frequently. This rule will not always apply, because some years are associated with moisture stress early in the season, and high rainfall later, but there will be a general tendency for early-maturing crops to be more tolerant of light soils and dry summers.

Table 7.3   Stage of growth of various cereal species on 12 May 1988 at
Gleadthorpe EHF

Species	Stage of growth (from Zadoks)	
Rye	55	half of ear emerged
Winter barley	49	first awns visible
Triticale	37	flag leaf just visible
Winter wheat	33	3rd node detectable
Oats	32	2nd node detectable

Source: Gleadthorpe EHF, ADAS.

The same is probably true when comparing irrigation responses of early- and late-drilled crops. One would expect a late-drilled crop with a shallow rooting system, and a later date of maturity, to be generally more sensitive to moisture stress and show a larger response to irrigation. In 1986, an experiment at Gleadthorpe demonstrated this very point with a larger irrigation response from a crop of spring barley drilled in March than from an earlier December-drilled crop. This will not always be the case, however. The early drought in 1988 gave a different result, possibly in part associated with the high level of lodging caused by irrigation of the late-drilled crop in that year.

While it is possible to argue a case for different responses to irrigation shown by rye, oats and the major cereals, I see no advantage at present in attempting to separate the results obtained with winter wheat, winter barley and spring barley. Until more experimental evidence is available, I shall group the results from these three crops in order to obtain a more reliable estimate of response to irrigation.

The low average response to irrigation of 0.34 t/ha from the nineteen experiments at Gleadthorpe demonstrates that, with 1989 prices, it is not possible to justify any level of investment in irrigation equipment if it is to be used solely on cereals. In fact there would have to be a very large shift in the balance between irrigation equipment cost and cereal prices before such an investment could be justified, and we have no reason to expect such a shift in the early 1990s.

There is another situation to examine, however. Many growers on light soils have an irrigation system for potatoes (and possibly sugar beet) which is not used until the beginning of June. They have therefore often taken the opportunity to irrigate cereals in May when the equipment is not being used for these crops. In this case the cereal response has only to cover the variable cost of irrigation to be worthwhile, as the fixed costs are covered by the root crops.

The value of such 'opportunity' irrigation will depend upon the balance between grain prices and the variable cost of applying water. In the 1980s we have observed a change in this balance, with a corresponding shift in the value of irrigating cereals. In the early 1980s, the balance between cost of application and grain price was such that irrigation was easy to justify on economic grounds. However, in the late 1980s, the value of grain has fallen, the variable costs of irrigation have increased and the situation is no longer quite so clear cut.

The total variable cost of irrigation varies from farm to farm, but close examination of the present circumstances at Gleadthorpe EHF shows that the variable cost of applying 25 mm on a hectare is £20 in 1989. Thus the total variable cost of applying the average 58 mm applied in the trials is around £46. If the price of cereals is taken as £100–£120 per tonne, the average response of 0.34 t/ha no longer covers the cost of irrigating.

This situation requires careful consideration. It is not possible to recommend that cereal irrigation be discontinued completely because, even at present costs, there are years such as 1984 when irrigation will still be very worthwhile. Indeed, in very dry years, irrigation is almost essential on the lighter soils, if reasonable yields are to be obtained.

In order to resolve this problem, it is necessary to find a middle course of action. Irrigating when the root zone capacity is depleted by 50 per cent will completely avoid drought stress, but it is too costly. Failing to irrigate at any time will risk major loss of yield and quality in very dry years. The sensible compromise is to irrigate only when the AWC is at risk of becoming severely depleted. This will not produce the maximum yield response in a dry year, but should provide an effective insurance against major drought losses.

Table 7.4 contains critical deficits for cereals based upon 50 per cent depletion of the root zone capacity, along with trigger deficits based upon 75 per cent depletion. These probably represent the two likely extremes, and the trigger deficit to be used should be within this range unless there are higher value crops on the farm requiring irrigation at the same time (see chapter 12 for a discussion of priorities). The actual choice of trigger deficit is a difficult one, because

*Colour plate 1
Automatic rainfall
shelter at
Gleadthorpe EHF.
A sensor operates
a motor that drives
the shelter over the
experiment when rain
is falling. The shelter
moves off the plots
when rain ceases*

*Colour plate 2
Aerial photograph of
the irrigation trials
at Gleadthorpe EHF.
A wide range of
crops is investigated
each year*

Max. Summer S.M.D.
Median value (mm).

	0-25
	26-50
	51-60
	61-85
	86-95
	96-105
	106-120

*Colour plate 3
Agroclimatic areas of
England and Wales,
based on summer
Soil Moisture Deficit*

Colour plate 4
Travelling raingun

Colour plate 5
Hosereel irrigator

Colour plate 6
Hosereel boom
irrigator

Colour plate 7   Linear move irrigator

Colour plate 8   Tensiometers

*Colour plate 9   Unirrigated
potatoes with high level of
common scab*

*Colour plate 10   Irrigated
potatoes with very low level of
common scab*

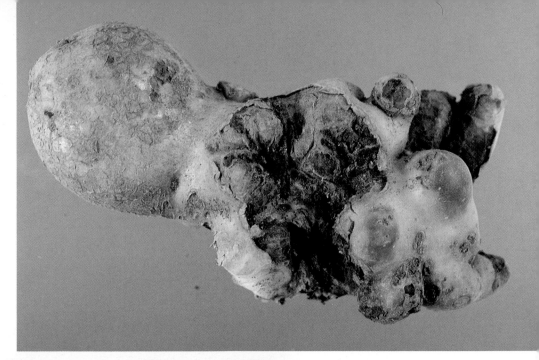

Colour plate 11
Powdery scab of potatoes

Colour plate 12
Blackleg of potatoes

Colour plate 13   Form of growth cracking characterised by cv Guardian in SCRI trials

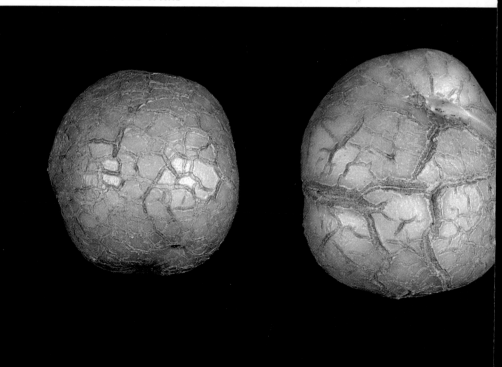

Colour plate 14   Form of growth cracking characterised by cv Record in SCRI trials

Colour plate 15   Dolls formed by temporary moisture stress

Colour plate 16   Extreme doll almost separated from the primary tuber

Colour plate 17   Unirrigated wheat at Gleadthorpe EHF in 1989

Colour plate 18   Irrigated wheat at Gleadthorpe EHF in 1989

Table 7.4   Critical soil moisture deficit for cereals according to soil texture

Soil texture	Winter cereals		Spring cereals	
	Trigger SMD (mm) based on 75% depletion of root zone capacity	Critical SMD (mm) based on 50% depletion of root zone capacity	Trigger SMD (mm) based on 75% depletion of root zone capacity	Critical SMD (mm) based on 50% depletion of root zone capacity
Sand	70	45	65	40
Loamy medium sand over sand	70	45	65	45
Loamy fine sand over sand	80	55	75	50
Medium sandy loam	125	80	120	80
Sandy clay or sandy clay loam	125	85	115	75
Clay or silty clay	120	80	105	70
Clay loam or silty clay loam	135	90	120	80
Fine sandy loam	145	100	135	90
Sandy silt loam	140	95	125	85
Silt loam	175	115	160	105
Acid or shallow peat	205	135	190	125
Deep fen peats	290	190	270	180

it is not simply related to soil type; it must take into account the price of grain as well as the cost of applying water. These can both be expected to undergo further change, and there is little point in offering recommendations here that are solely related to the economic situation in force at the time of writing. It is sufficient to say that winter cereals are likely to be more tolerant of higher deficits than spring cereals. Also, feed grain crops can be grown with somewhat higher SMDs than quality grain crops, as the market is likely to be more tolerant of drought defects in feed grain.

A further important consideration relates to the weather pattern that has been forecast. While it is almost impossible to forecast transpiration demand accurately, we do know that we can expect high rates when the weather is dry, hot and sunny, especially if there is an accompanying breeze. If conditions associated with high transpiration are expected, the prudent grower would irrigate as soon as the SMD approached the 50 per cent depletion stage. Conversely, if the weather is forecast to be cool and cloudy with the possibility of rain, it would be sensible to withhold irrigation until the SMD approached the higher end of the range. Such an attempt continually to link the trigger deficit to transpiration demand can be a very effective way of optimising irrigation on a low-value and moderately drought-tolerant crop such as cereals. However, as described in chapter 2, it does require continual monitoring of weather forecasts and the ability to respond to them in a flexible manner. This practice is not recommended for high-value drought-sensitive crops because the risk of being 'caught-out' is too high.

When to stop irrigating cereals

Experiments at Gleadthorpe have shown that irrigating cereals after flowering, particularly barley, will increase the risk of subsequent lodging. This is especially true if rainguns are used, and late irrigation is not advised. This means that all cereal crops run the risk of attaining high SMDs after flowering in a long dry spell, without the opportunity for irrigation. To lessen this risk, a final irrigation at the onset of flowering (or thereabouts) should be applied, even if the SMD is only approaching the lower end of the range shown in Table 7.4. If the weather then turns dry after flowering, the effects of drought during grain filling will be reduced. This tactic could also be used towards the end of May, if a grower has a large area of high-value crops such as potatoes to irrigate and is almost certain that he will not have the opportunity of irrigating cereals after the beginning of June.

Irrigation and cereal quality

Very little is known about the effect of irrigation on the different parameters of cereal quality. There are reports in the literature detailing the benefits of post-flowering irrigation on the quality of malting barley, by promoting grain filling and thus increasing specific weight and reducing grain nitrogen. The risk of lodging, however, from using commercial irrigation equipment post-flowering, is too high to justify an attempt at improving quality from irrigation at this time.

In recent years, staff at Gleadthorpe have attempted to monitor the effects on wheat quality of irrigation at different growth stages, but the spate of wet summers has led to a series of conflicting and confusing results, without any consistent trends. This study is continuing, along with another investigating malting quality of barley, and will hopefully offer some useful guidelines in the near future.

We know from experiments in 1976 that severe drought conditions are associated with poor grain quality of both wheat and barley, particularly as a result of poor grain filling and low specific weights. Such extreme cases of poor quality should, however, be avoided if the irrigation plans outlined in this chapter are adhered to.

At the beginning of chapter 1, I stated that it is because returns from irrigation in the UK are so marginal, that we have to be so painstaking in our approach. If we apply too little water, we will fail to obtain the benefits from irrigation. If we apply too much, the cost will be too great to make the returns worthwhile. Nowhere is this more apparent than in the cereal crop, simply because the returns from irrigating cereals are very marginal indeed in most years. It is only by the use of careful management that we can economically produce high grain yields of good quality on our lighter soils.

OILSEED RAPE

Very little experimental work has been conducted on the irrigation of oilseed rape, but fortunately the few experiments that have been done have given fairly consistent results.

Scientists in France have worked with both winter and spring oilseed rape. Their results showed that drought early in the season (before mid-May) had no effect on yield, but drought immediately before, during or after flowering could have harmful effects.

Similarly, scientists in Germany, working with winter oilseed rape, found that irrigation before flowering was unnecessary, but irrigation between flowering and 'green maturity' was beneficial on very light soils. They further declared that the water content of the soil during the critical phase should not be allowed to decline below 50 per cent of the root zone capacity. It must be noted, however, that these experiments were done in pots and did not involve farm-scale overhead irrigation.

During the period 1986–8, a series of experiments at Gleadthorpe investigated the use of irrigation on winter oilseed rape grown in UK conditions. The weather in 1986 consisted of a very wet April and May, followed by a dry period after flowering. Under these circumstances, irrigation after flowering (with the linear move irrigator) resulted in a yield increase of 0.32 t/ha. This result is in accord with the French and German work in showing that the crop is susceptible to drought in the period soon after flowering and irrigation can be of benefit then. In commercial practice, however, most irrigation in the UK is via mobile hosereel irrigators. Most oilseed rape crops are impenetrable at this time and the yield advantage from irrigating must be balanced against the damage caused to the crop. On balance, it must be concluded that irrigation with a hosereel after flowering should not be attempted, and only those growers with specialised equipment, such as centre-pivots or linear move irrigators, could really attempt late irrigation of the crop.

In 1987 the weather was quite the opposite. April and May were fairly dry, leading to a build-up of the SMD to 100 mm by the end of May, coinciding with the approaching end of flowering. From the beginning of June, rainfall was sufficient to prevent the crop coming under drought stress again. Under these conditions, irrigation before flowering and during flowering were examined, and both resulted in a yield decrease. The yield decrease from irrigation before flowering is in accord with the lack of response shown in both the French and German experiments at this growth stage. In this experiment, early irrigation increased vegetative growth, resulting in a considerably taller crop which gave the visual impression of benefit from irrigation. However, the yield at harvest was not improved, but showed a decrease of 0.3 t/ha.

Furthermore, irrigation during flowering also showed a similar decrease, but this is in contrast to the French and German work which showed an advantage from irrigation at this stage. The explanation may lie in the fact that the experiment at Gleadthorpe used overhead irrigation while the experiments abroad used trickle, and it is possible that overhead irrigation interfered with the fertilisation of the flowers.

The experiment in 1988 showed agreement with the two previous experiments in that irrigation during flowering decreased yield, possibly again because of interference with the fertilisation process. Irrigation during flowering also predisposed the crop to lodging subsequently. Irrigation after flowering again resulted in a yield increase, this time of the order of 0.45 t/ha, showing the advantage of late irrigation if suitable equipment is available.

In these experiments, the oil content of the seed was generally unaffected, except in 1988, when irrigation during flowering was associated with a slight decrease of around 1 per cent oil content.

In these trials a double-low variety (Ariana) was used. Glucosinolate levels were generally low, and unaffected by irrigation. Experiments in Australia have shown that drought stress can result in higher levels of glucosinolates, so it is possible that irrigation may have a useful effect on this character in a dry year.

In summary, the majority of growers with mobile hosereel irrigators are unlikely to obtain any benefit from irrigating oilseed rape unless SMDs become particularly high before flowering. The highest SMD attained pre-flowering in the Gleadthorpe trials was 50 mm, and this was not associated with a yield depression. It is likely that irrigation would be of benefit only if the SMD exceeded 75 mm at this time.

It is inadvisable to use overhead irrigation during flowering, and only those growers with suitable equipment, capable of irrigating without damage to the crop, should irrigate after flowering. A crop irrigated at this time would also be more likely to require a fungicide as the wet microclimate underneath the thick canopy would be an ideal environment for the development of disease caused by *Alternaria* and *Botrytis*.

DRIED PEAS

The irrigation requirements of dried peas are very similar to those of vining peas. Full details of how to irrigate peas can be found in chapter 8.

FIELD BEANS (*Vicia faba*)

There are many reported instances of the yield of field beans being limited by moisture stress, but there are few experiments that have critically examined the effect of different irrigation regimes on the crop. All experiments generally agree that field beans are sensitive to moisture stress at the time of flowering and during pod development, and irrigation at this time can lead to increased yields. Unfortunately, there is controversy regarding the effects of water stress before flowering, and there is a need for further experimentation on this subject. Under UK conditions, in most years, it is likely that irrigation from the onset of flowering until the pods are well filled will prevent the crop experiencing undue moisture stress.

As with cereals, it would be difficult to justify the capital costs of an irrigation system on field beans. It is likely, however, that the variable costs of irrigating can be easily covered if the guidelines below are followed.

Spring-sown beans

The advisability of irrigation depends very much on the lodging risk. General experience has shown that the short, strong-strawed early varieties show good responses to irrigation (up to 50 per cent yield increase). The yield responses from irrigation at Woburn Experimental Farm were greater than expected, by about 80 per cent. Day and Legg have reported that ten days of drought caused eighteen days' loss of growth.

Experience has also shown that later, weak-strawed taller varieties often show very little increase, if any, and should not be irrigated in most years. Results at Gleadthorpe EHF in 1988 are in accord with this view, and irrigation encouraged lodging in a late, tall variety, resulting in a significant loss of yield. These varieties should be irrigated at flowering and during pod-filling only if the height of the crop is restricted due to a lack of growth early in the season, the critical deficits presented here are exceeded, and there is no sign of rainfall expected. These conditions arose at Gleadthorpe in 1989, and a useful response to irrigation was obtained in that year. These results show the importance of correctly assessing lodging risk when irrigating spring sown beans.

Critical deficits for spring-sown field beans are presented in Table 7.5, based on 50 per cent depletion of the root zone capacity which, from root measurements at the Nottingham School of Agriculture, is assumed to consist of all available water in the top 40 cm, and easily available water in the next 30 cm. Some authors have argued that irrigation should be applied when the available water is depleted by 30 per cent, but this does not represent a conflict of opinion as they have calculated from the total amount of available water throughout the upper metre. Both methods should result in similar estimates of critical deficit.

Table 7.5   Critical soil moisture deficits for field beans according to soil texture

| | Critical SMD (mm) | |
Soil texture	Spring-sown	Winter-sown
Sand	30	35
Loamy medium sand over sand	30	35
Loamy fine sand over sand	40	45
Medium sandy loam	50	60
Sandy clay or sandy clay loam	50	60
Clay or silty clay	45	55
Clay loam or silty clay loam	50	60
Fine sandy loam	55	70
Sandy silt loam	55	65
Silt loam	65	80
Acid or shallow peats	75	95
Deep fen peats	105	135

There is experimental evidence to support the critical deficits presented in Table 7.5. French and Legg at Rothamsted have estimated the critical deficit to be 30 mm on a loamy sand over sand, and Hebblethwaite at the Nottingham School of Agriculture has shown the critical deficit to be approximately 65 mm on a sandy loam over clay.

However, French and Legg have also reported that the critical deficit was estimated to be over 80 mm on a silty clay loam over clay at Rothamsted, which is somewhat higher than the figure shown in Table 7.5. This relatively high critical deficit estimated at Rothamsted may be due to a particularly deep root system, as neutron probe work over many years has shown that roots extract water from 90 cm at this site, i.e. deeper than assumed here. Also, clay soils vary considerably in their water-holding capacity, well-structured subsoils having a much higher available water capacity than those with average structure (see Table 2.3).

The figures in Table 7.5 refer to soils of average structure. Allowance can be made for better or worse soil conditions by estimating critical deficits using the method described in chapter 2.

Autumn-sown beans (winter beans)

Penman showed that autumn-sown beans have a greater drought tolerance than spring-sown crops, probably as a result of a deeper root system. It is worthwhile irrigating them, however, when the risk of lodging is low. For autumn-sown beans, the high risk of lodging is associated with high plant populations. A trial at Gleadthorpe EHF in 1988 resulted in severe lodging and a yield reduction of nearly 1 t/ha when a crop with 29 plants/m^2 was irrigated. Crops with 15 plants/m^2 are likely to show a good response to irrigation. In fact, where irrigation can be guaranteed, the best strategy on light soils is probably to aim at a plant population of 12 plants/m^2. Critical deficits for winter beans are also presented in Table 7.5.

Chocolate spot and irrigation

Chocolate spot is the most important disease of field beans, particularly winter beans. It is caused by the fungi *Botrytis cinerea* and *B. fabae*. The disease can take the form of small red-brown spots or sometimes larger brown-black blotches. In severe epidemics plants are defoliated and even killed.

It is typically most severe after periods of wet weather, as it is encouraged by high relative humidity. Fitt *et al.* at Rothamsted showed that irrigation increased the levels of chocolate spot, but not markedly so in a dry year. Only when irrigation was applied in wet years did the severity of the disease show a significant increase. This is another argument in support of good irrigation scheduling and applying irrigation only when the SMDs approach the critical deficits.

REFERENCES AND FURTHER READING

DAY, W., and LEGG, B. J. (1981), 'Water relations and irrigation response', in P. D. Hebblethwaite (ed.), *The Faba Bean (Vicia faba L)* (Butterworth).

DAY, W., LEGG, B. J., FRENCH, B. K., JOHNSTON, A. E., LAWLOR, D. W. and JEFFERS, W. De C. (1978), 'A drought experiment using mobile shelters: the effect of drought on barley yield, water use and nutrient uptake', *Journal of Agricultural Science, Cambridge 91*, 599—623.

FITT, B. D. L., FINNEY, M. E. and CREIGHTON, N. F. (1986), 'Effects of irrigation and benomyl treatment on chocolate spot (*Botrytis fabae*) and yield of winter-sown field beans (*Vicia faba*)', *Journal of Agricultural Science Cambridge 106*, 307—12.

FRENCH, B. K. and LEGG, B. J. (1979), 'Rothamsted irrigation 1964—76', *Journal of Agricultural Science Cambridge 92*, 15—38.

GALES, K. and WILSON, N. J. (1981), 'Effect of water shortage on the yield of winter wheat', *Annals of Applied Biology 99*, 323—34.

HEBBLETHWAITE, P. D. (1982) 'The effects of water stress on the growth, development and yield of *Vicia faba* L', in G. Hawtin and C. Webb (eds), *Faba Bean Improvement* (ICARDA, Netherlands).

MORGAN, A. G. and RIGGS, T. J. (1981), 'Effects of drought on yield and on grain and malt characters in spring barley', *Journal of the Science of Food and Agriculture 32*, 339—46.

PREW, R. D. and GUTTERIDGE, R. J. (1988) *Friends of Rothamsted Newsletter No. 7*.

SALTER, P. J. and GOODE, J. E. (1967), 'Crop responses to water at different stages of growth', *Commonwealth Bureau of Horticulture, East Malling, Research Review No. 2*.

SELMAN, M. (1982), 'The effect of irrigation and nitrogen rate on the yield of certain cereals', *Experimental Husbandry 38*, 39—59.

WARD, J. T., BASFORD, W. D., HAWKINS, J. H., and HOLLIDAY, J. M. (1985) *Oilseed Rape* (Farming Press), 113—14.

Chapter 8

Irrigation of Field Vegetables

Irrigation can be important for vegetable crops because many are shallow rooting and therefore especially sensitive to water shortage. Also, vegetable production is a high-cost enterprise with a high-value end-product and many growers feel the need to have irrigation available as an insurance against drought.

There are several potential benefits from irrigating vegetables. The most obvious is a yield increase, but there are other important advantages such as improvement in plant establishment, continuity of supply to market or processor, and quality, including control of product size.

The irrigation requirements for quality, and growth throughout the season, vary between crops and need to be described under separate headings. Irrigation to aid establishment, however, involves principles that are common to most vegetables.

Irrigation to aid establishment of drilled crops

Many vegetable crops are grown in succession, and winter vegetables are usually sown in mid-summer, when the soil can be particularly dry in the upper layers, especially if SMDs have been allowed to build up under a preceding crop. Under these conditions, a small quantity of water (up to 20 mm) may prove beneficial to encourage uniform emergence and better early growth. This water is best applied a couple of days before drilling on soils that form a cap, but is better after drilling on the heavier soils.

Irrigation to aid establishment of transplants

It is important to wet transplants thoroughly before planting, particularly peat blocks. In dry soil conditions, however, this may not be sufficient and it is wise to keep the crop well supplied with moisture for as long as it takes for roots to grow out into the surrounding soil. Modules are particularly susceptible to drying out if planted so that they protrude above the soil surface.

It is safe practice to apply 10—12 mm water immediately after transplanting out, and repeat the application 3—4 days later. If the weather is particularly dry and windy at this time, it is possible that further applications will be required, but this is unlikely in most years.

Irrigation to aid harvesting

Lifting vegetables can be difficult in dry soil conditions, and irrigation may be used to soften the soil several days before harvesting. This is sometimes practised with crops such as beetroot, carrots and salad onions. Growers should be aware of the risks, however, and take note of the weather forecast at this time. If heavy rain follows irrigation, there could be obvious difficulties in harvesting from soil that is too wet.

Irrigation throughout the season or only at specific growth stages

Some vegetable crops respond to irrigation in dry soil conditions, regardless of the stage of growth, but others show particularly sensitive stages, e.g. peas at flowering, brussel sprouts when the lower buttons are 15—18 mm diameter, etc. The requirements for vegetables are extremely varied, so those for each crop will now be described separately.

VINING PEAS

Most of our knowledge about the irrigation of peas came from a series of experiments carried out at NVRS, Wellesbourne during the 1960s. Peas are very responsive to irrigation, but the timing of irrigation is crucial for success with this crop.

Irrigation before flowering

The experiments at NVRS showed that irrigation before flowering was associated with a marked increase in haulm growth. In these experiments, on a sandy loam soil, this increase amounted to about 60 per cent. Unfortunately, however, irrigation at this time actually resulted in a slight decrease in the yield of peas. It was therefore concluded that, in most years, irrigation should not be applied during this period. There are three notable exceptions to this rule:

1. If the seedbed is very dry, it is possible that germination could be reduced. Irrigation may be of benefit in aiding germination under these conditions. It should be noted that this is a rare situation because, in most seasons, the soil is likely to be near field capacity at the time of planting and irrigation may produce a waterlogged seedbed and consequent poor establishment as a result.
2. Some short varieties may produce too little haulm in dry conditions, and extra haulm growth could make mechanical harvesting easier.
3. In the unusual conditions of a severe drought the crop may become so severely wilted that yield would be subsequently affected. In this rare event, irrigation should be applied. It is difficult to state precisely the level of SMD that should trigger irrigation during this period, but I would suggest that it be based on 80 per cent depletion of the root zone capacity (see Table 8.1).

Irrigation at the start of flowering

This is the time when the greatest benefits from irrigation are obtained. In the experiments at NVRS, irrigation at this time resulted in an average yield increase of 30 per cent. This yield resulted from an increase in the number of pods per plant and the number of peas per pod. In such situations, when the opportunity for irrigation is confined to a short period, it is not advisable to wait until the root zone capacity is depleted by 50 per cent. If the SMD at this time approaches 40 per cent depletion of the root zone capacity, 25 mm of irrigation should be applied.

According to Salter and Drew, there is a marked reduction in root growth at the start of flowering. Up until this stage, the roots continually grow and extend into fresh soil moisture reserves but, at the start of flowering, they almost stop growing and access to water becomes restricted. Water stress can then occur quickly unless there is an adequate renewed supply from rainfall or irrigation.

The reduction of root activity at the time of flowering is a characteristic of many annual plants and is possibly responsible, in part at least, for the fact that so many of them appear to be particularly sensitive to moisture stress at this time.

Irrigation during petal-fall

Irrigation is not advised during this period. In the experiments at NVRS, irrigation at petal-fall did not result in any yield increase, possibly due to the fact that root growth recommences and fresh soil moisture reserves become available. Indeed, irrigation at this stage could be detrimental, as the wet petals may stick to the pods and encourage the growth of *Botrytis*.

Irrigation during pod-swelling

Irrigation applied when vining pea crops show a tenderometer reading of around 80 has resulted in yield increases averaging 10–20 per cent. It is interesting to note that root growth is interrupted again at this time. On very light soils, 25 mm water can be applied, but this may create harvesting difficulties on the heavier soil types, where only 12 mm should be applied.

Irrigation at this stage will delay the maturing of vining peas by approximately two days. This is not a problem as the effect is consistent and predictable, and can be taken into account when vining.

In summary, the irrigation of peas is a very complex subject compared with some other crops, because it must be linked closely to the stage of crop growth. However, if the proper attention is given to the irrigation of this crop, it can result in very good yield responses, with a financial value exceeding that of many of our other crops.

Table 8.1   Suggested critical soil moisture deficits for peas according to growth stage and soil texture

| | Critical SMD (mm) | | |
Soil texture	Before flowering	Start of flowering	Pod-fill
Sand	45	20	20
Loamy medium sand over sand	45	25	25
Loamy fine sand over sand	55	30	30
Medium sandy loam	75	35	35
Sandy clay or sandy clay loam	75	35	35
Clay or silty clay	65	30	30
Clay loam or silty clay loam	75	35	35
Fine sandy loam	85	40	40
Sandy silt loam	80	40	40
Silt loam	95	50	50
Acid or shallow peats	115	60	60
Deep fen peats	165	80	80

PEAS FOR HAND-PICKING

The critical deficits used for vining peas can also be used for hand-picked peas. However, vining peas have a short flowering period, and most of the pods mature at the same time, allowing a single harvest at the correct stage of maturity. With such varieties, it is fairly easy to pinpoint the stage of growth at which irrigation should be applied. Some varieties used for hand-picking may flower over longer periods of time, and the pods mature not simultaneously, but in a succession over a period. With such varieties, it is often difficult to time irrigation correctly. In these circumstances, it is probably best to apply 25 mm water at the first sign of flowering, another 12 mm when the most advanced pods are just beginning to swell, and a further 12 mm after the first pick has been completed.

Mangetout peas have increased in popularity in recent years. There is no experimental evidence available to show how best to irrigate them, but they also mature in succession over a period of time, rather than simultaneously. I would therefore advise irrigating them in this same manner.

BEANS

Runner beans (*Phaseolus coccineas*)

A study in the 1960s showed that the annual yield of the commercial runner bean crop in England and Wales was positively correlated with the total March–October rainfall. In contrast, the yields of the closely related green bean (*Phaseolus vulgaris*) and of the pea crop showed no clear correlation. This infers that runner beans are particularly sensitive to soil moisture stress and, further, that their sensitivity is not confined to narrow growth stages as is found with peas and green beans.

A likely explanation for this is that runner beans have an indeterminate habit, flowering and pod-setting over a long period throughout the summer. As flowering is known to be a particularly susceptible stage for moisture stress in legumes, one may expect to find that the crop responds to water throughout this long period.

In contrast, peas and green beans flower and set pods within a relatively short time, and so are only particularly susceptible to moisture stress for a short period. They would therefore be less likely to show a good correlation between yield and total summer rainfall.

In early experiments at Wellesbourne on a sandy loam soil, yield responses ranging from 14–68 per cent were obtained as a result of irrigating runner beans from the early green flower-bud stage, i.e. when the greenish tight clusters of flower-buds could be clearly observed in the leaf axils. More flowers opened on irrigated plots and these gave rise to more and heavier pods than those harvested from unirrigated plots. Irrigation applied at an earlier stage, when the crop had three expanded leaves, had no effect on yield.

There is no detailed evidence on the critical deficits for runner beans, but I would suggest that good responses could be obtained if the SMD were not allowed to exceed the levels presented in the right-hand column of Table 8.1 (critical deficits for peas) after the early green flower-bud stage. Some advisers advocate the use of slightly lower critical deficits, and they may be correct, but there is no experimental evidence to support this view. In conditions of extreme drought, I would irrigate before this stage, as recommended with peas (see previous section).

Many gardening books recommend that spraying water onto runner beans and green beans during flowering will increase the proportion of flowers that actually set pods. This practice was investigated at NVRS, Wellesbourne, many years ago and the evidence does not support it. First, the proportion of flowers that set pods in both a wet and dry year was found to be similar (about 30 per cent), although pod yields were four times higher in the wet year. Second, various spraying treatments were applied to plants in a dry year. These included applying water through a mist nozzle attached to a hand-held hose, moistening flowers with a fine syringe, and even comparing rain-water with a bore-hole supply. These dampening treatments made very little difference, although a slight decrease in pod-setting was sometimes recorded. It was concluded that dampening the flowers will result in a yield improvement if enough water is applied, not because of an increase in pod-setting, but because sufficient water falls to the soil to act as light irrigation and increases the total number of flowers that open. Flower spraying can, in fact, deter the natural insect pollinators that are essential to crop production.

Green beans (*Phaseolus vulgaris*)

Green beans (sometimes called dwarf or French beans) are quite shallow rooting and sensitive to moisture stress at certain growth stages. The optimum stages for applying irrigation are quite similar to those for peas. Early irrigation will increase haulm growth, sometimes improving mechanical harvesting, but will not increase yields *except in conditions of severe drought*. Irrigation at the green bud stage is beneficial, and growers have been advised to apply 25 mm water if the SMD is 25 mm or more, regardless of soil type. As with peas, the crop should not be irrigated throughout the flowering stage as the incidence of *Botrytis* may be increased by wet petals sticking to the developing pods. Irrigation at the stage of pod development will probably increase yield. In fact, some advisers believe this to be the best time to irrigate green beans, if only one application is possible. Late irrigation should be avoided as it could make travel difficult for the harvester.

Broad beans

Broad bean varieties fall into two categories, viz. those with an indeterminate flowering habit, with pods filling at different times along the stem, and those varieties which have pods that mature simultaneously in clusters at the tips of the stems. The latter types show two distinct stages of growth which are particularly moisture-sensitive. The best time to irrigate is at flowering, and 25 mm water should be applied if the SMD is above 25 mm, regardless of soil type. Irrigation will also increase yields if applied at pod-set onwards, using the critical deficits given for spring-sown field beans (see Table 7.5). With indeterminate varieties, these stages of growth overlap and irrigation is usually applied at any time from early flowering.

Growers should be aware that irrigation can be associated with an increased risk of chocolate spot infection.

SPRING-SOWN BULB ONIONS

Until recent years, there was very little experimental work on the irrigation requirements of onions. There has been an increased level of research in the last ten years, but the requirements are still not fully understood. However, the crop is generally regarded as having a low priority for irrigation, and only marginally economic responses are expected.

Irrigation to aid establishment

On some soils, a combination of rain followed by drying conditions can lead to the formation of a dry soil crust or cap on the surface. This cap can form a mechanical barrier to the emergence of the seedling, resulting in low plant populations. In extreme situations, the crop may benefit from *small* amounts of water applied in order to wet and soften the crust, but this should be considered only as a last resort. The amount applied should not exceed 12 mm, as higher amounts risk saturating the soil around the seed and replacing one type of stress with another. Large droplets should be avoided where possible as they may damage any seedlings that have already emerged ahead of the majority. Finally, if the weather remains dry, the crust may form again, and it could become necessary to repeat the procedure.

In recent years, there has been an increase in the use of soil-conditioners as a band-spray after drilling, to prevent surface capping, and this should eliminate the need for irrigation at this stage.

In 1987, staff at Gleadthorpe investigated the need to irrigate transplants at the time of planting into sandy soils. The soil was wet (field capacity) at the time of planting, but no rain followed during the whole of the subsequent month. Under these conditions, there was no benefit obtained from irrigating the crop with 5 mm water at two days after planting, followed by 10 mm water at eight days after planting. In 1988, there was adequate rainfall after transplanting, and it was not possible to repeat the investigation. However, there is a general opinion held by experienced growers and advisers that onion transplants should be irrigated in dry periods on light soils, and the practice is advisable, provided that care is taken not to damage the small plants with large heavy drops.

Irrigation for bulking

Irrigation to increase bulb size and yield has been investigated at several sites during recent years. The results are shown in Table 8.2, along with summer rainfall (May—August).

Table 8.2   Yield response from irrigation of spring-sown bulb onions

Site	Year	Summer rainfall (mm)	Yield response (t/ha)
Arthur Rickwood EHF	1976	108	28.0
Stockbridge House	1984	115	16.5
Oak Park	1984	120	21.7
Luddington EHS	1981	196	4.3
Arthur Rickwood EHF	1977	208	0
Luddington EHS	1981	223	10.0
Gleadthorpe EHF	1985	232	0
Gleadthorpe EHF	1988	238	0
Oak Park	1982	244	6.5
Gleadthorpe EHF	1986	268	2.1
Luddington EHS	1982	280	0
Gleadthorpe EHF	1987	285	0

Source: Various sources.

The average yield response is 7.4 t/ha, but this covers a tremendous range, resulting mainly from variation in summer rainfall, as can be seen in Table 8.2. It is obvious that irrigation is likely to give a large benefit in dry years, when the summer rainfall is less than 150 mm, but a very unpredictable result when summer rainfall is greater than this.

Critical deficits for high financial return crops are normally based on 50 per cent depletion of the root zone capacity, while those for low financial return crops such as cereals allow for a greater amount of depletion. How should onions be treated? Two sets of critical deficits for onions, based on 50 per cent and 80 per cent depletion respectively, are given in Table 8.3. Where maximum responses are required in a dry year the deficits based on 50 per cent depletion can be used, but these will inevitably involve a degree of over-watering in a normal year. If the deficits based on 80 per cent depletion are used, over-watering in a normal or wet year will be minimised, but there will be a somewhat lower response in a dry year. Each grower must choose his irrigation policy based on the expected value of the crop, the irrigation requirements of other crops on the farm, and the expected frequency of dry summers (less than 150 mm May–August rainfall). At Gleadthorpe, the expected frequency of such summers is only four years out of thirty, and it could be argued that this is infrequent enough to justify the use of the lower critical deficit, and the higher deficit would be suitable. On the other hand, in some eastern counties, the frequency could be much higher and use of the lower critical deficit would be advisable.

Table 8.3 Critical soil moisture deficits for onions according to soil texture

| | Critical SMD (mm) | |
| | Based on 50% depletion of root zone capacity | Based on 80% depletion of root zone capacity |
Soil texture		
Sand	25	40
Loamy medium sand over sand	25	40
Loamy fine sand over sand	35	55
Medium sandy loam	40	65
Sandy clay or sandy clay loam	40	60
Fine sandy loam	45	70
Sandy silt loam	45	70
Silt loam	55	85
Acid or shallow peat	60	95
Deep fen peats	90	145

In between these two extremes, intermediate figures could be used. A grower should be aware of the frequency of such dry summers occurring in his locality before deciding on the optimum irrigation plan for his onion crops.

Irrigation and quality

In most of the experiments listed in Table 8.2, irrigation had no measurable effect upon the quality of the onions. There were two exceptions to this, however.

At Oak Park, in 1984, it was reported that the visual quality of the irrigated onions after drying and grading was superior to the unirrigated crop.

At Arthur Rickwood EHF, in 1976, there was slight evidence of increased rotting and sprouting in store from irrigated plots but, in spite of this, 97 per cent were still sound on 26 March. Differences in dry matter were more marked, being about 2 per cent lower on the irrigated plots.

In general, taking all the trials into account, irrigation had little or no effect on quality.

Nitrogen for the irrigated crop

During the period 1985—8, the level of nitrogen for irrigated onions on sandy soils after cereals was investigated at Gleadthorpe. Results showed that the optimum amount is 130 kg/ha.

With their shallow root system, onions are particularly vulnerable to nitrogen loss as a result of leaching. A split application is therefore recommended on sandy soils, with 40 kg/ha applied in the seedbed and 90 kg/ha applied approximately a month later.

Even with a split application, the crop is susceptible to leaching losses in a wet year. The 1987 trial received high rainfall in June (150 mm) and showed a response to a further late 'top-up' dressing. In other years, this extra was not required and, indeed, may have had detrimental effects.

It is important to apply the correct amount of nitrogen. If too little is applied, yields will be low. If too much is applied, thick-necked bulbs can result. In years of high spring or summer rainfall, it is recommended that growers seek advice from ADAS regarding the need for late 'top-up' dressings.

Less nitrogen is required on other soils. An application of 90 kg/ha is generally sufficient on all other soils except fen peats, for which 30 kg/ha is sufficient.

AUTUMN-SOWN BULB ONIONS

Irrigation to aid establishment

It is important to ensure that autumn-drilled crops germinate and emerge quickly so that the optimum plant size is achieved before winter. If the soil is particularly dry at the time of drilling, as can often occur with a crop that is drilled in August, satisfactory and timely establishment may not be achieved. This may be especially true if the previous crop has extracted most of the available water from the upper few inches of the soil. In these conditions, small quantities of water may prove beneficial to encourage uniform emergence and early growth. This water may be applied several days before drilling in light soils, but is better applied after drilling on medium and heavy soils.

Irrigation for bulking

There is very little experimental evidence on this subject and some advisers have suggested that the critical deficits shown in Table 8.3 could be used for autumn-sown onions, but irrigation is generally considered to be of low importance except in very dry conditions.

SALAD ONIONS

Little experimental work has been done on the response of salad onions to irrigation. The following guidelines are based upon general experience.

Timely establishment is important, in order to give a continuity of supply. If the soil is dry, water can be applied before or after sowing to get crops established.

After the crop is established, irrigation may be given to increase growth, keeping the SMD below 25 mm on all soil types. This is likely to be particularly beneficial on crops sown before mid-July, but is also necessary for later-sown crops in some years.

Irrigation is often applied at harvest too, as it aids lifting if the soil is dry.

TURNIPS

Stanhill at NVRS irrigated turnips at various stages and measured the yield response. He found that irrigation up until ten days after sowing was beneficial to the crop in dry conditions. However, he also reported that irrigation between ten and eighteen days after sowing resulted in a yield depression at harvest. This was then followed by a third period which extended until harvest, and was associated with a large yield response to irrigation.

Guidelines of critical deficits for turnips are given in Table 8.4. Stanhill reported that the maximum yield was obtained when the crop was returned to field capacity each day. Irrigating whenever the root zone capacity was depleted by 50 per cent was insufficient for maximum growth and resulted in an 18 per cent loss of yield from moisture stress. The effects of this moisture stress were greater if it occurred later in the season. For this reason, the critical deficits for the latter part of the season, presented in Table 8.4, are based on 30 per cent depletion of the root zone capacity, and not 50 per cent. There is, however, a need for fresh experimental evidence to examine critically whether such a cautious approach is strictly necessary, particularly as Stanhill reported elsewhere that the rate of evapotranspiration was not reduced until 60 per cent of the available water had been used.

The quality of cooked turnips has been shown to be influenced by irrigation applied in the four weeks preceding harvest. Irrigation during this stage gave the roots a milder flavour and smoother texture.

Table 8.4   Critical soil moisture deficits for turnips according to soil texture and stage of growth

| | Critical SMD (mm) | |
Soil texture	3—8 weeks after emergence	From 8 weeks after emergence
Sand	15	20
Loamy medium sand over sand	25	25
Loamy fine sand over sand	25	25
Medium sandy loam	25	40
Sandy clay or sandy clay loam	25	40
Fine sandy loam	30	45
Sandy silt loam	25	45
Silt loam	35	55
Acid or shallow peat	40	60
Deep fen peats	55	85

CARROTS

Trials with carrots between 1964 and 1969 at Gleadthorpe showed little benefit from irrigation. Only in 1967 was there a response, when irrigation increased yield by 5 t/ha. In recent years the subject has been studied again at Gleadthorpe, and the results are shown in Table 8.5.

Table 8.5   Yield response from irrigation of carrots at Gleadthorpe EHF

Year	May–August rainfall (mm)	Irrigation applied (mm)	Yield response (t/ha)
1985	232	86	22.3
1986	268	59	8.4
1987	285	71	0.0
1988	238	136	16.2

Source: Gleadthorpe EHF, ADAS.

None of the four years can be classed as being dry, but the average yield response was quite high. It is difficult to explain the contrast between the results of these recent experiments and those in 1964–9, or even between years within this series, especially as the early period also contained some years when the May–August rainfall was less than was recorded in this recent trial series.

Nevertheless, the results from recent trials show that the irrigation of carrots can produce high responses. It should be noted that these experiments were harvested in October, and represent a late crop. Where carrots are produced for the earlier market, the yield responses will certainly be much less.

Quality

An experiment at Wellesbourne in 1956 showed that very intensive irrigation (returning to field capacity every day) produced a greater yield of edible roots, but they were less sweet, had weaker 'carrot' flavour, less colour and rougher texture than unirrigated carrots. Even more remarkable was the fact that these effects were also detectable when the irrigation was restricted and applied during a two-week period only, beginning twelve weeks before harvest. I am not aware of any recent studies of this subject in the UK but there have been a series of experiments conducted in Norway. Riley and Dragland at the Kise Agricultural Research Station obtained large increases in yield from irrigation, but found that water applied during the three weeks before harvest produced a less sweet taste.

Some workers have associated splitting with irrigation, but the relationship is not entirely clear. It is probable that if large SMDs are allowed to develop before irrigation is applied, splitting may result.

It is also unclear whether there is a connection between irrigation and cavity spot, and a trial series to examine this is under way at Gleadthorpe.

An irrigation plan for carrots

For some years, ADAS has advised that carrots should not be irrigated between sowing and the 4 true-leaf stage. Experiments in Norway have supported this view, and workers there have suggested that irrigation at this time retards growth by creating lower soil temperatures. Further evidence against irrigation at this time has been obtained recently at Gleadthorpe. The results shown in Table 8.5 were obtained by irrigating only after the 6 true-leaf stage. Irrigating earlier gave no further yield benefit although, unlike the Norwegian experiments, it did not depress final yield.

Where high seed rates are used, as for the production of carrots for 'canning', the crop may require a somewhat different treatment and irrigation after sowing could be advisable in dry conditions. In an experiment at Wellesbourne, in 1963, seven different plant densities were used ranging from approximately 50 to 850 plants per square metre. At the lower plant densities, irrigation during the seven weeks following germination depressed the yield of marketable roots, but at higher plant populations early irrigation increased yield.

Critical deficits for carrots have not yet been determined experimentally. However, there is circumstantial evidence to suggest that their rooting pattern is similar to that of turnips. It is also a high-value crop, capable of producing very good responses to irrigation. Taking all this into account, it seems wise to use a critical deficit of 30 per cent depletion of the root zone capacity, as with turnips, until there is good evidence to suggest otherwise. After reaching the 6 true-leaf stage, carrots should be irrigated before reaching the critical deficits suggested in Table 8.6.

BEETROOT, PARSNIPS AND SWEDES

Very little experimental irrigation has been carried out on these crops. Irrigation was generally believed to be only marginally economic, if at all, on parsnips and swedes. For this reason, irrigation was not usually advised in the past, except when soil conditions were exceptionally dry at sowing. It is my view, however, and that of my colleagues in ADAS, that these crops will all respond to irrigation, especially beetroot and parsnips grown at high densities for small root

Table 8.6 Critical soil moisture deficits for carrots according to soil texture

Soil texture	Critical SMD (mm)
Sand	20
Loamy medium sand over sand	25
Loamy fine sand over sand	25
Medium sandy loam	40
Sandy clay or sandy clay loam	40
Fine sandy loam	45
Sandy silt loam	45
Silt loam	55
Acid or shallow peat	60
Deep fen peats	85

production on light soils. The recommended critical deficits are about 25 per cent higher than those given for carrots in Table 8.6, e.g. 25 mm on a sand, etc.

BRASSICAS

Cauliflowers

In experiments at NVRS during the period 1955–60, Salter showed that early summer cauliflowers were very sensitive to drought stress and required frequent irrigation during the seedling stage of growth and subsequently. Moisture stress also affects the quality of the produce: browning discolouration of the curd has been shown at Luddington EHS to be related to a water shortage in the plant while the curd is developing.

In Salter's experiments, watering whenever the available water in the root zone was depleted by 50 per cent did not produce maximum yield, which was only obtained if the soil was kept at field capacity by daily watering, or if the depletion of the available water in the root zone was limited to 25 per cent. This requires very intensive irrigation, especially when the root system is small, and Salter recognised that it was impractical in most commercial situations. Critical deficits based on 25 per cent depletion are presented in Table 8.7 for those few growers who have the capability to water frequently with such small amounts, especially if the weather is warm and sunny, and the potential transpiration rates are high. Most commercial growers, however, are unlikely to irrigate until the SMD exceeds 25 mm, although some irrigate their transplants at an SMD of 20 mm during the first month after planting.

Salter's work demonstrated two further practical implications. Under conditions of shortage of water, labour or equipment, it may not be possible to irrigate intensively for maximum yield. In this case, provided there is enough water for the crop to establish properly, the most economic return for a reduced level of irrigation was obtained by delaying the application until the later stages. This was substantiated by work during 1962–4 at Luddington EHS where, in two out of the three years, a single application of 25 mm when half the plants had curds about 30 mm in diameter gave as high or almost as high a yield as watering throughout life at 25 mm SMD.

Salter also showed a strong link between irrigation response and plant populations. In the absence of irrigation, yields decreased if the spacing was closer than 60 × 60 cm. With irrigation, however, a higher yield could be obtained at a closer spacing, depending on the size of curd required. It is now well established that intensive irrigation will enable plants for early summer production to be more closely spaced than normal.

Since the early 1980s, there has been a marked change in the husbandry of cauliflowers, with much of the crop now being grown as *module* transplants at high populations. The higher sensitivity of such crops to moisture stress has been partly responsible for the fact that the crop is now largely confined to the better moisture-retentive soils. The critical deficits in Table 8.7 will still apply, but there is now some uncertainty about the best time to apply limited water. Some advisers believe that it may be better to apply this water during vegetative development in order to build the necessary framework of leaves before curd initiation, but this still needs experimental verification.

Cabbages

A major effect of irrigation is usually to increase the leaf growth of plants. With a crop such as cabbages, which consists mainly of leaves, it is therefore to be expected that irrigation will have a marked effect on yield. Experiments by Drew on a sandy loam soil of the Newport series at NVRS, Wellesbourne in the early 1960s supported this view. One experiment had five irrigation treatments, consisting of restoring the soil to field capacity whenever the SMD reached 12 mm, 25 mm, 50 mm or 75 mm, and an unirrigated control. The highest yield was obtained from irrigation at 12 and 25 mm deficit, and yields progressively decreased when using the higher deficits (see Table 8.8).

Table 8.7   Critical soil moisture deficits for cauliflowers and cabbages according to soil texture and stage of growth (see text)

| | | Critical SMD (mm) | | |
| | Drilled crop | Until 5 weeks after emergence | Until 9 weeks after emergence | Subsequently |
Soil texture	Transplanted crop	1 week after transplanting	3 weeks after transplanting	Subsequently
Sand		5	10	15
Loamy medium sand over sand		5	15	15
Loamy fine sand over sand		10	15	20
Medium sandy loam		10	20	30
Sandy clay or sandy clay loam		10	20	30
Fine sandy loam		10	25	30
Sandy silt loam		10	20	30
Silt loam		10	30	40
Acid or shallow peat		10	35	45
Deep fen peats		15	50	65

Table 8.8   Effect of different water regimes on the yield of summer cabbage grown on a medium sandy loam

SMD (mm) at which water was applied	Yield as a percentage of intensively irrigated treatment
12	100
25	97
50	88
75	78
No irrigation	48

Source: Drew, National Vegetable Research Station.

This result confirms that cabbages are particularly sensitive to moisture stress and the critical SMDs given for cauliflowers should be used for this crop.

This experiment again confirms that, as with cauliflowers, the later stages of growth are particularly important for irrigation. The SMD of 50 mm first occurred only three weeks before harvesting, and the SMD of 75 mm was not reached until ten days before harvesting, yet irrigating at these times gave considerable yield increases over the unirrigated plots. The explanation probably lies in the growth pattern of the crop. Cabbages grow very slowly in their early life, and then show very rapid growth shortly before they are harvested. As a result, moisture stress during this final phase is likely to have a greater effect. This experiment, taken along with the results of other experiments by Drew, led him to conclude that when the amount of water available for irrigation is limited, it should be reserved for use during the last three weeks before harvest, when the crop is particularly susceptible to moisture shortage.

Brussels sprouts

As with many vegetable crops, irrigation may be required to aid the establishment of drilled and transplanted crops. A description of this was given earlier in the section on onions.

There is no evidence on which to estimate a set of critical deficits for sprouts, but research has shown that the crop can benefit from a single application of 40 mm when the lower buttons are 15−18 mm in diameter, if at that time the SMD exceeds 40 mm. Personally, I would always prefer to use two applications of 25 mm during this period, rather than 40 mm in one application.

LETTUCES

There has been no experimental work conducted to determine the irrigation requirements of iceberg lettuces, but there have been many experiments with cabbage lettuces. Unfortunately, the results from these trials are inconclusive and conflicting. At Luddington EHS and Efford EHS, the crop did not respond to irrigation if it was planted in soil which was close to field capacity. Further, in wet years, yield was decreased as a result of irrigation. At NVRS, Salter found in one series of experiments that the highest yield was obtained if the SMD was kept below 12 mm at all times, but in another series, irrigation had no effect. In another experiment, Winter at NVRS reported that the best result was obtained with an initial wet period followed by dry conditions after the beginning of hearting.

The most recent work is that of Sale in 1963 and 1964 at NVRS, Wellesbourne. This work, on a medium sandy loam, showed that plants irrigated whenever the SMD reached 30 mm had the same fresh weight as those irrigated whenever the SMD reached 18 mm, although both were considerably heavier than unirrigated plants. Interestingly, plants which received a single irrigation eight weeks after planting, returning the soil to field capacity, were very nearly as large as those in the other two treatments. A similar result was obtained both for the drilled crop and for transplants raised under glass and then planted in the field.

The work of Sale showed that lettuces gain weight very slowly during the initial period, and there is then a rapid increase in growth during a three-week period beginning about eight weeks after sowing. Both ground cover and the development of soil moisture deficits followed this same pattern. Seven weeks after sowing, the ground cover was only 30 per cent and the SMD was only 25 mm; after eleven weeks, the ground cover was over 90 per cent, and the SMD had reached 75 mm. As described in the section on cabbages, it is this final period that is most important, and irrigation during this time is likely to show the highest responses.

In these experiments, irrigation produced a marked improvement in the quality of the crop, and over 95 per cent of the plants were marketable. Conversely, the unirrigated plants had a dark green colour and a tough, wrinkled appearance, with only 68 per cent reaching a marketable condition.

From these results we can conclude that the best results will be obtained by irrigating the crop well and keeping the SMD below the critical deficits given as a guideline in Table 8.9. If water is likely to be in short supply, however, a single application of 25 mm should be given three weeks before cutting.

Table 8.9   Critical soil moisture deficits for lettuces according to soil texture and stage of growth

| Soil texture | Critical SMD (mm) | |
	6 weeks after emergence	8 weeks after emergence
Sand	15	20
Loamy medium sand over sand	15	25
Loamy fine sand over sand	20	30
Medium sandy loam	20	35
Sandy clay or sandy clay loam	20	35
Fine sandy loam	20	45
Sandy silt loam	20	40
Silt loam	25	50
Acid or shallow peat	30	60
Deep fen peat	40	85

Summary

In chapter 2, a simple method of estimating critical soil moisture deficits was described. It was also stated that most crops will avoid stress if the root zone capacity does not deplete by more than 50 per cent. This is true for a range of crops including potatoes, sugar beet, cereals, etc. However, the needs of some vegetable crops appear to be more complex, sometimes owing to the fact that often only a small part of the plant is harvested, and total growth of the plant is not necessarily correlated to this desirable fraction (e.g. peas). With many vegetables, there is only a small window of time when irrigation can be applied profitably. Also, as with potatoes, the visual appearance of the product is often affected by moisture stress, and this quality factor can have more effect than yield in determining final profitability. The many factors involved in producing high yields, high quality and a continuity of supply of field vegetables necessitate that the particular requirements of each crop should be understood thoroughly before applying irrigation. The best results will only be obtained when the irrigation is tailored to the specific needs of crop and soil, and planned with a good irrigation scheduling technique.

References and Further Reading

BLACKWALL, F. L. C. (1969), 'Effects of weather, irrigation, and pod-removal on the setting of pods and the marketable yield of runner beans (*Phaseolus multiflorus*)', *Journal of Horticultural Science 44*, 371–84.

BLACKWALL, F. L. C. (1971), 'Pod-setting and yield in the runner bean (*Phaseolus multiflorus*)', *Journal of the Royal Horticultural Society 96*, 121–30.

BREWSTER, J. L. (1977), 'The physiology of the onion', *Horticultural Abstracts 47*, 17–23.

DRAGLAND, S. (1984), 'Effects of drought at different growth stages of vegetables', *Proceedings of the North-Western European Irrigation Conference*, Billund, Denmark, 136–9.

DREW, D. J. (1966), 'Irrigation studies on summer cabbage', *Journal of Horticultural Science 41*, 103–14.

MAFF (1982), 'Irrigation', *Reference Book 138*.

RICE, B., JELLEY, M. and PHELAN, P. (1982), 'Irrigation of sugar beet and other crops', *An Foras Taluntais*, Annual Research Report of Plant Sciences and Crop Husbandry, Oak Park Research Centre, 49–50.

RICE, B., JELLEY, M. and PHELAN, P. (1984), 'Irrigation of sugar beet and other crops', *An Foras Taluntais*, Annual Research Report of Plant Sciences and Crop Husbandry, Oak Park Research Centre, 48–51.

RILEY, H. and DRAGLAND, S. (1988), 'Irrigating field vegetables for quality 2. Effects of drought at different stages of growth', *Irrigation News 14*, 19–27.

SALE, P. J. M. (1966), 'The response of summer lettuce to irrigation and plant spacing', *Journal of Horticultural Science 41*, 31–42.

SALE, P. J. M. (1966), 'The response of summer lettuce to irrigation at different stages of growth', *Journal of Horticultural Science 41*, 43–52.

SALTER, P. J. (1959), 'The effect of different irrigation treatments on the growth and yield of early summer cauliflower', *Journal of Horticultural Science 34*, 23–31.

SALTER, P. J. (1960), 'The effect of different soil moisture conditions during seedling stage on the growth and yield of early summer cauliflower', *Journal of Horticultural Science 35*, 239–48.

SALTER, P. J. (1960), 'Irrigation of early summer cauliflower', *Agriculture 67*, 59–62.

SALTER, P. J. (1961), 'The irrigation of early summer cauliflower in relation to stage of growth, plant spacing and nitrogen level', *Journal of Horticultural Science 36*, 241–53.

SALTER, P. J. (1962), 'Some responses of peas to irrigation at different growth stages', *Journal of Horticultural Science 37*, 141–9.

SALTER, P. J. (1963), 'The effects of dry or wet soil conditions at different growth stages on the components of yield of a pea crop', *Journal of Horticultural Science 38*, 321–34.

SALTER, P. J. (1963), 'The right time to water peas', *Grower 59*, 859.

SALTER, P. J. and DREW, D. H. (1965), 'Root growth as a factor in the response of *Pisum sativum L.* to irrigation', *Nature 206*, 1,063–4.

SALTER, P. J. and GOODE, J. E. (1967), 'Crop responses to water at different stages of growth', *Commonwealth Bureau of Horticulture, East Malling, Research Review No. 2*.

SALTER, P. J. and WILLIAMS, J. B. (1963), 'Irrigation of carrots', *Annual Report of the National Vegetable Research Station*, 48.

SALTER, P. J. and WILLIAMS, J. B. (1967), 'The effect of irrigation on pea crops grown at different plant densities', *Journal of Horticultural Science 42*, 59–66.

STANHILL, G. (1956), 'Irrigation of carrots', *Annual Report of the National Vegetable Research Station*, 47.

STANHILL, G. (1958), 'Effects of soil moisture on the yield and quality of early turnips. I. Response to different sustained soil moisture regimes', *Journal of Horticultural Science 33*, 108–18.

STANHILL, G. (1958), 'Effects of soil moisture on the yield and quality of early turnips. II. Response at different growth stages', *Journal of Horticultural Science 33*, 264–74.

WINTER, E. J. (1968), 'Irrigation of summer cauliflower – experiments at Efford, Luddington and Wellesbourne', *Experimental Horticulture 18*, 52–9.

Chapter 9

Irrigation of Fruit

There has been much experimental work to show that timely irrigation of fruit crops can help achieve higher and more consistent yields of good-quality fruit. All fruit crops are adversely affected by moisture stress during the period of fruit development, leading to reduced size and quality. Further, moisture stress can lead to reduced vegetative growth, which may limit yield in subsequent years.

Before describing the benefits that have been obtained with each crop, some general observations regarding irrigation scheduling are necessary.

The use of the Penman method for estimating evapotranspiration is less reliable for fruit crops than for those crops discussed in previous chapters. There are several reasons for this.

Estimation of crop cover in fruit crops is difficult. If viewed from above, it is apparent that fruit crops, with their wide row spacings, generally have a low crop cover (typically 50 per cent of incoming solar radiation is intercepted by tree row crops). In most top fruit orchards, trees are grown with grass (or sometimes strawberries) in the tractor alleyways, with a resulting increase in net water usage. In this case a Penman-type water balance equation gives a reasonable estimate of total water usage, and is suitable for irrigation scheduling, providing the crop cover of both the trees and the grass is accounted for.

Many UK soft fruit crops, however, are grown with an overall herbicide system of management to eliminate all weeds and grass that would compete for water. Yet, even in this case, it is clearly wrong to estimate evapotranspiration, and SMD build-up, on the basis of low crop cover because water extraction will be occurring at full rate in the immediate vicinity of the tree or bush where the majority of roots are located.

Many fruit crops are irrigated with trickle systems and various types of micro-spitters and mini-sprinklers which only apply water close to the individual plants. When using such systems, it is clearly incorrect to assume that the irrigation amounts are evenly spread over the total soil surface, so a conventional approach to calculating water balance may be wrong.

146

Consider the situation carefully. We are not attempting to produce an estimation of SMD throughout the field, but we should only be estimating SMD in the relevant areas from which the plants are extracting their water. Thus a typical crop cover pattern for apples with overall herbicide should be considered as 5 per cent early in the season, and 10 per cent at full bloom, but 70 per cent about thirty days after full bloom. The moisture status of the soil in the wide spaces between plant rows is not relevant if this moisture is not available to the crop.

With some fruit crops, however, it is difficult to define precisely the relevant areas, especially if trickle irrigation is used. It is common to find that the area of ground exploited by roots, the area covered by foliage, and the area actually being watered by the irrigation system are all widely different. In such situations, it is not yet precisely clear how a water balance should be constructed.

Crops such as strawberries are sometimes grown under polythene, either as a complete covering or as a mulch on the soil surface. Evapotranspiration in these crops is not understood thoroughly. (The same problem is also encountered with early potato production in some areas.) There is a further problem with estimating the penetration of rain into the soil and around the root system. Some growers use overhead irrigation with the polythene in place, assuming that much of the water channelled into the spaces between the covered beds will move laterally within reach of the roots. We have very little information at present as to what proportion of the applied water becomes available in this way, and an attempt at a water balance calculation is likely to contain errors.

The simplest way of overcoming these problems is to schedule irrigation by using a direct form of soil moisture measurement, such as tensiometers. However, ADAS workers have attempted to produce and test soil moisture models that cope with these situations, and they are now included in Irriguide. At the time of writing, critical experimental testing has not been completed, but the results so far obtained in commercial practice have been very satisfactory. Due to the varied nature of the rooting patterns reported in the literature, however, I have not attempted to relate critical SMDs to an allowable depletion of the root zone capacity, as has been done for other crops in this book. J. Atwood of ADAS has suggested some critical SMDs that are successfully used for fruit in commercial practice, and these form the basis for some of the tables presented in this chapter.

BLACKCURRANTS

Experiments conducted thirty years ago at East Malling Research Station (now the Institute of Horticultural Research (IHR) East Malling) showed a significant yield increase from irrigating blackcurrants. The application of 45 mm water when the SMD was at 45—60 mm has produced yield increases of 50 per cent in dry seasons at East Malling, where the soil is a fine sandy loam.

At Luddington EHS, where blackcurrants were grown on a coarse sandy soil, the average yield over six years was increased by 50 per cent when irrigation was used. This was achieved by applying 50 mm whenever the SMD reached 75 mm, using a sprinkler system.

A further study at Luddington of trickle irrigation produced yield increases of 75—225 per cent in the particularly dry seasons of the mid-1970s (see Table 9.1).

Table 9.1   Irrigation of blackcurrants at Luddington EHS

Treatment	*1974*	*1975*	*1976*
		Yield (t/ha)	
Trickle irrigated	11.0	19.6	13.3
Unirrigated	6.3	10.3	4.8
% increase	76	90	221

Source: ADAS.

More recent irrigation trials at Luddington EHS between 1980 and 1982 have again shown yield responses varying between 0 and 54 per cent, depending on the weather. The size of individual berries was also improved.

These trials all show that irrigation can improve the yield and size of fruit if applied in a dry summer, but it has also been found that irrigation will increase shoot growth and crop potential for the following year.

Norwegian experiments have produced guidelines for those growers using tensiometers to schedule irrigation on blackcurrants. Irrigation was applied at various levels of soil tension extending from 50 to 250 centibars (measured with electrical resistance blocks). The highest yield was obtained from irrigating before the tension exceeded 50 centibars, but this was not significantly greater than irrigating at 70 centibars, whereas higher tensions produced significantly lower yields.

Where a form of water balance calculation is used for scheduling,

the critical SMDs shown in Table 9.2 have been found to work successfully in commercial practice. Only in exceptional circumstances will
irrigation be required before the end of flowering. The evapotranspiration for April is typically 50—60 mm, and the crop cover will
usually increase from almost nothing to about 50 per cent at the end
of April. The critical SMD of 50 mm will not often be exceeded.

Table 9.2 Critical soil moisture deficits for blackcurrants according
to soil texture

| | Critical SMD (mm) | |
| | Before end of flowering (usually April) | After end of flowering (usually May—July) |
Soil texture		
Sand	50	35
Loamy medium sand over sand	50	35
Loamy fine sand over sand	50	35
Medium sandy loam	75	50
Sandy clay or sandy clay loam	75	50
Clay or silty clay	75	50
Clay loam or silty clay loam	75	50
Fine sandy loam	75	50
Sandy silt loam	75	50
Silty loam	75	50
Acid or shallow peats	75	50
Deep fen peats	75	50

Source: ADAS.

RASPBERRIES

In 1975—6, irrigation experiments with raspberries at Luddington EHS,
using trickle irrigation during the 3—4 week period from the com-
mencement of ripening, showed the exceptionally large responses that
are possible in dry years. In contrast, in the wet summer of 1974,
no yield response was obtained (see Table 9.3).

The lack of any marketable yield in 1975, the fruit being small and
leathery, shows the necessity of irrigation in some situations if good
yields are to be produced every year.

Experiments at East Malling between 1957 and 1962 also confirmed
that raspberries produce good responses to irrigation. During the three
years 1957 to 1959, the extra yield resulting from irrigation was 28,
42 and 35 per cent respectively. The irrigation treatments consisted

Table 9.3   Irrigation of raspberries at Luddington EHS

| Treatment | Yield of marketable fruit (t/ha) | | |
	1974	1975	1976
Malling Jewel			
Irrigated	14.3	10.3	11.8
Unirrigated	14.1	0	4.8
Glen Clova			
Irrigated	17.8	8.5	16.1
Unirrigated	17.8	0	8.8

Source: ADAS.

of irrigating whenever the root zone capacity was depleted by 40—60 per cent, up until the end of August. This increased yield of berries was also associated with an increase in cane production and growth, however, which was not always beneficial. Excessive cane growth can shade the fruit and delay ripening, as well as creating difficulties in harvesting. Too much cane growth may also result in many berries being chafed, particularly in windy conditions, and this can result in a high level of *Botrytis* infection.

The workers concluded that irrigation is best restricted to the pre-picking period, which should increase fruit yield without excessive cane growth. They concluded that a single application of 50 mm is probably sufficient in most years, provided it is given when the berries are about to grow rapidly, i.e. when they show the first signs of colour change due to ripening (usually in mid-June). This strategy was tested in later experiments, with resulting yield increases of 20—35 per cent in 1961—2.

D. Mackerron of the Scottish Crop Research Institute investigated the effect of irrigation on raspberries over the period 1968—77, with a somewhat different result. He showed that a single irrigation of water just prior to fruit ripening produced a smaller response compared with irrigation throughout the season whenever the soil moisture tension reached 40 centibars (see Table 9.4).

In commercial practice, growers often irrigate intensively during the period of ripening and picking, but will also irrigate before this period if conditions are particularly dry. The critical soil moisture deficits are shown in Table 9.5. Irrigation after picking is unlikely to be beneficial, except in establishing young plantations where extra growth is required.

Table 9.4 Yield of raspberries (t/ha) under different irrigation regimes

| | | Treatment | |
| | | Irrigated at pink | Irrigated throughout |
Year	Unirrigated	fruit stage only	season
1968	2.2	2.1	2.5
1969	9.2	9.3	10.5
1970	4.9	4.2	4.5
1971	6.2	6.3	7.4
1974	12.6	13.7	13.6
1975	8.1	7.7	9.8
1976	12.7	13.3	12.6
1977	10.5	11.1	13.4
Mean	8.3	8.5	9.3

Source: Scottish Crop Research Institute.

Table 9.5 Critical soil moisture deficits for raspberries according to soil texture

| | Critical SMD (mm) | |
| | Before | During ripening |
Soil texture	ripening	and picking
Sand	50	30
Loamy medium sand over sand	50	30
Loamy fine sand over sand	50	30
Medium sandy loam	50	35
Sandy clay or sandy clay loam	50	35
Clay or silty clay	50	35
Clay loam or silty clay loam	50	35
Fine sandy loam	60	35
Sandy silt loam	60	35
Silt loam	75	35
Acid or shallow peats	75	35
Deep fen peats	75	35

Source: ADAS.

STRAWBERRIES

It has been established for many years that irrigation of fruiting strawberries will generally result in increased yield and size of fruit. This
was sometimes associated with increased *Botrytis* infection, but the
benefits from irrigation usually outweighed any losses from disease.

Early experiments at East Malling Research Station showed that,
on a medium loam of 75–90 cm in depth, a potential SMD of 100 mm

was associated with plants visibly suffering from drought. Irrigation was said to be 'urgently necessary' and produced a yield increase of 2 t/ha of marketable fruit. Staff at East Malling have estimated from long-term weather records that these conditions have occurred about one year in two for July-harvested varieties. It is likely that irrigation would also increase yields under less severe conditions, and some benefit can be expected more frequently. Varieties harvested in late August–September are likely to show a considerable benefit nearly every year, as the potential SMD is regularly about 100 mm at the end of August.

The most comprehensive set of experiments in the UK was conducted by Hughes at Efford EHS during the period 1960–3, on a fine sandy silt loam. The best yields were obtained by regular irrigation of 25 mm every time the SMD reached 50 mm, until picking ended. In the dry year of 1962, this gave a yield response of 6.8 t/ha (35 per cent) with one-year-old plants, and 5.8 t/ha (28 per cent) with two-year-old plants, compared with the unirrigated plants. Responses in the other years were smaller or non-existent. Single irrigations applied at certain times (e.g. picking) gave smaller responses.

Irrigation had no discernible effect on the amount of fruit damaged by slugs, straw, pickers or other hazards, but did result in increased levels of *Botrytis*. This increase was small, however, compared with the effect of seasonal conditions on the incidence of the disease (see Table 9.6).

Table 9.6   Effect of irrigation on the percentage of *Botrytis*-infected strawberries

		Unirrigated	Irrigated with 25 mm at 50 mm SMD
One-year plants	1960	20	19
	1961	8	11
	1962	4	13
Two-year plants	1961	14	21
	1962	4	9
	1963	35	30

Source: Efford EHS.

The irrigated plots were of a superior quality, the berries being bright and glossy, while the fruits from unirrigated plots were described as dull, dry and unattractive looking. The produce was also studied to determine whether irrigated fruit deteriorated more rapidly, but no differences were found. Travel tests were also conducted, and showed that the stage of ripeness at picking and packing determined quality on arrival, with no detrimental effect from irrigation.

According to Doorenbos and Pruitt, in their FAO paper on guidelines for predicting crop water requirements, strawberries are very shallow rooting. They recommended that, for the purposes of calculating irrigation requirement, and hence critical deficits, the effective rooting depth should be taken as 30 cm, which is the shallowest of all the forty-five crops considered by them. This does not appear to be borne out by experience with the crop in the United Kingdom, however. In fact, Higgs and Jones at IHR East Malling have studied the rooting pattern of strawberries, and recorded approximately 50 per cent of the roots in the top 20 cm, 35 per cent in the 20–60 cm layer, and 15 per cent below 60 cm. They also showed that strawberries extract a significant amount of water from depths below 30 cm. There is no doubt that rooting depth will vary between sites, but this work demonstrates that under some conditions strawberries can root quite deeply.

Higgs and Jones also compared soil moisture deficits calculated on the basis of the modified Penman method against direct measurements using neutron probe data. They concluded that the modified Penman method was accurate enough for agronomic purposes, and this is also the conclusion from ADAS experiences with Irriguide. The method is explained in chapter 4.

Suggested critical soil moisture deficits for strawberries are presented in Table 9.7, where it can be seen that they are linked to growth stage of the crop. During the development of the fruit, the crop is particularly sensitive to moisture stress, and again at the end of the season when the following year's flowers are initiating (September for June-bearers, later for everbearers). During these stages, the critical deficits have been set at a low level to ensure no drought stress. However, at other stages, when the crop is less sensitive to moisture levels, the critical deficits have been arbitrarily increased. These deficits have not been subjected to experimental testing, but they are being used at present in commercial practice and have achieved good results. They are based on the assumption that irrigation is applied with a raingun, as is commonly the case. Where trickle irrigation is used, it is likely that growers will irrigate more intensively during fruiting, picking and flower initiation on sands and loamy sands, keeping the SMD below 20 mm.

The results of strawberry irrigation experiments described here were obtained using traditional growing techniques. In recent years, there has been considerable development in new cultural techniques such as polythene covers on raised beds and waiting-bed plants to extend the season. Irrigation plays a vitally important role in these techniques.

Table 9.7   Critical soil moisture deficits for strawberries according to
soil texture and stage of growth

| | Critical SMD (mm) | | | |
Soil texture	Early season	Fruiting and picking	After fruiting	During flower initiation
Sand	50	30	50	30
Loamy medium sand over sand	50	30	50	30
Loamy fine sand over sand	50	30	50	30
Medium sandy loam	50	30	50	30
Sandy clay or sandy clay loam	50	30	50	30
Clay or silty clay	50	30	50	30
Clay loam or silty clay loam	50	30	50	30
Fine sandy loam	60	40	60	40
Sandy silt loam	60	40	60	40
Silt loam	75	50	75	50
Acid or shallow peats	75	50	75	50
Deep fen peats	75	50	75	50

Source: ADAS.

Polythene-covered raised beds

An increasing proportion of strawberries are now being grown in
polythene-covered raised beds. As the polythene reduces the amount
of rain actually falling on the crop, there is a greater likelihood of
plants suffering from moisture stress. Irrigation is particularly important even at those times of year when there would normally be sufficient
rainfall to meet plant needs. Similarly, overhead irrigation with the
polythene in place can result in much water loss through run-off from
the polythene, the amount depending on whether the polythene is
perforated and on the size of the perforations.

Strawberries grown in this way can be irrigated much more efficiently with trickle irrigation. As described in chapter 4, and again
at the beginning of this chapter, tensiometers are more likely to give
a reliable result in this situation than is a water balance sheet.

Extension of season

One way of doing this is by using waiting-bed plants, which are
produced especially for this purpose. They are planted in May or June
and require rapid establishment in order to flower and fruit within sixty
days, thus producing a crop during the last week of July and the first
two weeks of August. Irrigation is particularly important in aiding rapid
establishment and growth.

Sometimes waiting-bed plants fail to achieve their full potential, even though there is a plentiful supply of moisture in the soil. The foliage often grows relatively faster than the root system, and the plants simply do not have enough roots to keep up with transpiration demand at this time of year. They therefore suffer from moisture stress, which can be overcome only by 'misting'. A fuller description of this technique can be found towards the end of this chapter.

APPLES

The first significant series of experiments designed to investigate the benefits of irrigation on apple trees was conducted at East Malling over the period 1953–60. The variety used was Laxton's Superb and the trees were eight years old at the beginning of the study. Two soil types were used for the study, viz. a loamy fine sand over loamy sand (10 per cent AWC) and a deep loamy fine sand (14 per cent AWC). Treatments consisted of watering whenever the soil moisture tension at 30 cm depth reached 13, 27 or 67 centibars, together with an unirrigated treatment. The soil moisture release characteristics of both soils were fairly similar over this range of tensions, and these irrigated treatments equate to SMDs of 22, 40 and 55 mm.

The results showed that irrigation had a marked effect on the growth of trees and the marketable yield of apples. The best treatment consisted of irrigating at an SMD of 40 mm, and showed a 55 per cent increase in marketable yield of apples over the whole period compared with the unirrigated trees. The treatment involving very intensive irrigation (at a deficit of 22 mm) only showed a 44 per cent yield increase, which the workers explained as being due to nutrient leaching. They identified potassium, phosphate and magnesium as being leached from the upper soil layers, but only potash became limiting.

The third treatment, which was the least intensive and consisted of irrigating at an SMD of 55 mm, produced very little response in the initial years of the study, when trees were small and root growth was still fairly restricted. In later years, when the trees had increased in size and vigour, and produced a better root system, this treatment produced growth and yield equal to irrigating at 40 mm. This is an excellent demonstration of the need to match the choice of critical deficit against root growth of the crop.

The extra yield came from a combination of increased number and increased size of fruit. Other than this effect on size, there were no differences in quality between irrigated and unirrigated apples; colour, taste and occurrence of defects were all assessed and found to be similar between treatments.

Workers at Luddington EHS conducted a similar trial series on the variety Cox's Orange Pippin during the period 1958–66. The soil was a deep, coarse sandy loam with an average AWC of 13 per cent. The four experimental treatments consisted of watering when the SMD reached 25, 75 and 125 mm, along with an unirrigated control. However, the workers reported that the calculations used to assess the SMD probably produced overestimates by as much as 20 per cent, and the actual deficits at which water was applied were probably lower than intended. In the initial years of the experiment, irrigation responses were quite high, with the three irrigated treatments producing yields that were 42, 37 and 25 per cent more, respectively, than the unirrigated controls. Irrigating at an SMD of 75 mm, as calculated, was the most economical treatment.

The timing of irrigation was investigated at East Malling in a series of experiments over the period 1967–75, on a fine sandy loam soil with an AWC of 15 per cent, using trees that had been planted in 1961. Treatments consisted of: irrigation in June only (early), irrigation from July to mid-September (late), early and late irrigation, and an unirrigated treatment. The results are shown in Table 9.8.

Table 9.8   Effects of irrigation timing on yield of apples

	Irrigation			
	None	Early	Late	Early + late
Numbers of picked apples	1,001	1,088	1,019	1,042
Mean fruit size (g)	104	103	120	115
Weight of fruit (kg/tree) in highest grade (colour, size and skin quality)	206	260	344	340

Source: Goode, Higgs and Hyrycz, East Malling.

The results clearly show the benefit from late irrigation alone, with no added benefit gained from combining early and late irrigation. In fact, early irrigation can cause too many fruits to be retained, thus reducing fruit size subsequently. Also, early irrigation promotes excessive shoot growth. From these experiments we can conclude that, with mature trees, there is little point in irrigating before the end of June unless SMDs reach particularly high levels, say, 100 mm. Young trees less than three years old, however, are unlikely to be as tolerant of dry soils early in the season and it is recommended that they be irrigated throughout.

This same study also provided evidence on the effect of irrigation on rooting. Contrary to some earlier suspicions that irrigation inhibited rooting (see chapter 6 for a fuller discussion on this subject) there was a general increase in the quantity of fibrous roots of irrigated trees (see Table 9.9).

Table 9.9   Quantity of fibrous roots <1 mm diameter (mg per tree) in a sample volume of 0.0134 m³ (each value is the mean of 32 samples)

Depth (cm)	Unirrigated	Irrigated
0–15	25	34
15–30	22	23
30–45	7	12
45–60	4	8

Cultivar: Cox's Orange Pippin on MM.104

Source: Goode, Higgs and Hyrycz, East Malling.

These trials also showed an improvement in skin quality from trees that were irrigated late. Rough russet was always worse on fruit from trees that had not received late irrigation. Skin-cracking was severe in two years of the experiment which were associated with high water stress in August or September. In these years, the trees receiving early irrigation showed symptoms equally as severe as those on unirrigated trees. However, the disorder was almost completely absent from trees receiving late irrigation.

Another form of skin-cracking was identified in 1973, and also shown to be related to irrigation. Fruit harvested with blemish-free skins developed fine cracking after a period of storage, and the loss of marketable yield from this disorder was then considerable. The results are shown in Table 9.10 and the benefits from late irrigation are readily apparent.

The year of the study (1973) contained an exceptionally dry August (only 13 mm rainfall) and trees not receiving irrigation at that time showed marked drought effects (leaf yellowing and defoliation), along with a reduction in the rate of fruit growth. This dry period was followed by a period of heavy rainfall (126 mm) in September. The workers concluded from the study that the cracking was a consequence of different growth rates of the inner and outer tissues of the fruits resulting from fluctuating water supply conditions. This appears to be a similar phenomenon to the growth-cracking observed in the potato crop under similar circumstances. However, in this case the fruit

Table 9.10   **The percentage of trees with fruit having obvious symptoms of skin-cracking when removed from store in December**

Irrigation	Nitrogen rate kg/ha			
	26	78	130	182
None	38	81	81	88
Early	56	88	69	94
Late	0	0	0	0
Early + late	0	6	0	0

Source: Goode, Fuller and Hyrycz, East Malling.

appeared to be in good condition at the time of picking in October, and the cracks only appeared after storage.

Late irrigation has also been shown to reduce the development of bitter pit in cold store, possibly due to an effect of water supply on the mineral composition of the fruit. Finally, fruit from unirrigated trees and trees receiving early irrigation developed more soft roots in store compared with trees receiving late irrigation, and this is probably associated with the incidence of fine cracking already described.

From the results of this comprehensive set of experiments and considerable practical experience, J. Atwood of ADAS has produced irrigation schedules that are appropriate for achieving high yields and high-quality fruit, and these schedules are being used with success in commercial orchards today (see Table 9.11). Some adjustment (up to around 25 per cent) should be made where rootstocks are known to be particularly shallow or deep rooting. However, it must again be emphasised that there are difficulties involved in calculating SMDs under apple trees because of the complex relationship between crop cover, rooting zone, wetted area, etc.

The advantages of using tensiometers with fruit crops have already been discussed. Workers at East Malling have successfully irrigated apple trees by using pairs of tensiometers, placed at depths of 30 and 60 cm, and at a distance of 75 cm from the trees. Irrigation is applied to keep the soil moisture tension below 66 centibars and 13 centibars at the 30 and 60 cm depths respectively.

PEARS

Pears are generally believed to produce a greater response to irrigation than is obtained with apples. Critical SMDs for pears are presented in Table 9.12.

Table 9.11 Critical soil moisture deficits for apples according to soil texture

| | Critical SMD (mm) | | |
| | Mature trees | | Young trees |
Soil texture	April May June	July August early September	(less than 3 years old) All months
Sand	100	40	30
Loamy medium sand over sand	100	40	30
Loamy fine sand over sand	100	40	30
Medium sandy loam	100	70	40
Sandy clay or sandy clay loam	100	70	40
Clay or silty clay	100	70	40
Clay loam or silty clay loam	100	70	40
Fine sandy loam	100	85	40
Sandy silt loam	100	85	40
Silt loam	100	100	40
Acid or shallow peat	100	100	40
Deep fen peats	100	100	40

Source: ADAS.

Table 9.12 Critical soil moisture deficits for pears according to soil texture

| | Critical SMD (mm) | | |
| | Mature trees | | Young trees |
Soil texture	April May June	July August early September	(less than 3 years old) All months
Sand	100	30	30
Loamy medium sand over sand	100	30	30
Loamy fine sand over sand	100	30	30
Medium sandy loam	100	55	40
Sandy clay or sandy clay loam	100	55	40
Clay or silty clay	100	55	40
Clay loam or silty clay loam	100	55	40
Fine sandy loam	100	65	40
Sandy silt loam	100	65	40
Silt loam	100	80	40
Acid or shallow peat	100	80	40
Deep fen peats	100	80	40

Source: ADAS.

PLUMS

The guidelines given for the irrigation of apple trees have also been used successfully to irrigate plums in commercial orchards, except that the switch to the intensive regime will usually occur two weeks earlier, i.e. in the middle of June.

'Gumming' of plums is associated with an interrupted water supply. A gum is produced around the stone, and subsequently oozes out through a small hole in the skin, creating a sticky surface which can become dirty. Well-irrigated plums should avoid this disorder.

CHERRIES

Moisture supply is a particularly important requisite of quality in cherries. If the supply is temporarily interrupted, they suffer from splitting. In fact, this is such a serious problem that cherries are not usually grown on soils with a low AWC.

Cherries can also be irrigated according to the guidelines given for apples, shown in Table 9.11.

WATER STRESS CONTROL BY MISTING

In 1979, Goode, Higgs and Hyrycz at East Malling published the results of a series of tests on the use of irrigation 'mist' to control water stress in apples. It is firmly established that, under conditions of high transpiration (e.g. at midday in hot, sunny weather), the total transpiration loss from a plant is greater than its uptake. Even under UK conditions, plant performance is sometimes reduced by moisture stress which cannot be overcome by soil application of water. A promising approach is to reduce stress by applying water directly to leaf surfaces. While a leaf is wet there will be negligible transpiration flow through the plant. In their tests, they applied irrigation 'mist' to the foliage for 1½ minutes every 30 minutes between 11 a.m. and 3.30 p.m. or, on days of prolonged stress, until 4.30 p.m. The misting was not applied on dull, cloudy and cool days, or when it was raining.

The technique was very successful, producing higher yields as a result of an increase in fruit numbers, and higher quality, because of considerably reduced levels of bitter pit, severe russet and fine cracking. As the benefits from irrigation applied to the soil resulted from increased fruit size, in dry seasons the benefits from overhead misting and soil irrigation were additive, and a combination of both treatments produced the best yields of prime grade fruit (see Table 9.13).

Table 9.13   Effect of misting and irrigation on fruit set and cropping of apple trees

	Treatment			
	None	Irrigation	Misting	Irrigation + misting
Mean number of fruits per tree per year (1974–6)	358	358	416	482
Mean fruit size (g) (1974–6)	99	118	105	111
Mean yield (kg per tree) of graded fruit per year (1973–6)	30	39	40	46

Source: Goode, Higgs and Hyrycz, East Malling.

If this technique is to be used efficiently and economically, it is necessary to have specialist control equipment to monitor temperatures and to operate the system. With such equipment, it should be possible to automate the sprinklers so they are switched on for, say, 2 minutes every 30 minutes between 10 a.m. and 4 p.m. when the temperature is greater than 18°C in sunny conditions. A further refinement involves using the same equipment to keep the soil moisture deficit low by applying a heavy mist when the deficit is high or increasing, but a light mist when the deficit is low.

This technique has also been used successfully on strawberries at East Malling. In an experiment in 1984, waiting-bed plants were irrigated twice daily in order to maintain the soil at field capacity. This was then supplemented by 'misting' with mini-sprinklers positioned 50 cm above the soil surface, operated automatically for 2 minutes in every 30 minutes, between 10 a.m. to 4 p.m. The best treatment consisted of 'misting' for the first 32 days after planting, and produced an increase of 75 per cent in the yield of quality fruit. There were varietal interactions, however, and the variety Karona gave a better yield if misting was continued for the first 42 days after planting.

At the time of writing, there are very few growers who have adopted the technique, because of the higher capital costs involved, scepticism as to whether the benefits at IHR could be repeated in commercial orchards, and a fear of increased disease levels. It remains to be seen whether this relatively new technique will expand in the future.

FROST PROTECTION BY OVERHEAD SPRINKLER IRRIGATION

Overhead sprinkler irrigation can be used to control frost damage on developing buds or flowers, and many growers use it, particularly for blackcurrants and, to a lesser extent, apples. Frost will damage buds if the air temperature drops below a critical level which is dependent on the type of crop, and the stage of flower or fruit development. Although this critical temperature is variable, most deciduous fruit species are only damaged at temperatures below $-2°C$. Frost protection hinges on the principle that water sprayed on the buds can be allowed to freeze and, provided the ice is kept 'warm' at $0°C$, there will be no damage from frost. The ice is maintained at this temperature by continually spraying it with water.

In scientific terms, as the irrigation water freezes into ice, it releases a certain amount of heat energy which is transferred to the bud or flower. This heat is known as the latent heat of fusion. Provided water is applied at a sufficient rate, and enough ice is produced, the heat released will be sufficient to keep the temperature of the bud around freezing point. However, if sprinkling is interrupted for more than a few minutes once the trees or bushes are wet, all the water may freeze and the flowers or buds are then likely to suffer more damage than they would if no sprinkling had been done at all.

The rate of application required for frost protection is dependent on temperature, wind speed and other meteorological variables, as well as the size of the bud and the amount of ice already accrued on the buds during the course of a frost. As an example, consider the effect of wind speed. An increase in wind speed will result in some of the applied water being lost by evaporation, and this will have the unwanted opposite effect of cooling the buds. To overcome this heat loss, extra water is required to freeze and produce heat. Thus the rate of application should ideally take the wind speed into account. In fact, if wind speeds are in excess of 5 mph, frost protection by means of spray irrigation may not be practical. With high wind speeds the evaporative losses are so great that, because there is a practical limit to the amount of water that can be applied, irrigation may have the undesirable net effect of actually cooling the buds and increasing frost damage.

In commercial practice, most sprinkler systems are designed to apply water at a fixed rate estimated to be necessary to protect against the worst conditions of frost normally encountered. As a guideline, $3-3.5$ mm/hour should provide protection down to $-5°C$, in the absence of wind. Inevitably this results in unnecessarily high application rates for much of the time, with a considerable amount of run-off leading to very wet soils. A good drainage system is essential, as several inches of water may be applied over a few consecutive days. Typically for blackcurrants, the sprinklers switch on as the air temperature drops to 0°C, and switch off as it rises from below freezing to 0°C and the ice starts to melt. For apples it is possible to delay starting until the air temperature has dropped to −1°C, thus saving water, but again sprinkling should continue until the air temperature increases to 0°C. Suitable alarm systems are available to warn the operator when the air temperature is approaching 0°C, or alternatively, automatic operating systems may be used. Whichever the case, it is important that the temperature sensors be placed immediately *outside* the protected area, but in a position equivalent to the coldest area to be protected. Also, the sensors should be protected from early morning sunlight.

Hamer at East Malling Research Station has conducted a thorough study of frost protection methods and has produced a computer model to estimate the rates of application required under different conditions. He concludes that a sprinkler system capable of varying application rate according to conditions would have distinct advantages, as adequate protection against frost could then be obtained with greatly reduced amounts of water. Unfortunately, it is not possible to reduce the application rate by reducing pressure, as this would result in unacceptably poor distribution patterns with low uniformity. He recommends that sprinklers be continually switched on and off during conditions of low demand with the mode of application (on, off or pulsing) set according to air temperature recorded with sensors immediately outside the protected area. Such systems are now being successfully used in New Zealand, and have good potential for use in the UK.

REFERENCES AND FURTHER READING

BAILEY, J. (1986), 'Irrigating top and soft fruit', *Irrigation News 10*, 50—4.

COLE, I. (1987), 'Irrigation of fruit, 1. Top fruit', *Irrigation News 11*, 13—17.

DOORENBOS, J. and PRUITT, W. O. (1977), 'Crop water requirements', *FAO Irrigation and Drainage Paper No. 24*, 88.

GOODE, J. E., FULLER, M. M. and HYRYCZ, K. J. (1975), 'Skin-cracking of Cox's Orange Pippin apples in relation to water stress', *Journal of Horticultural Science 50*, 265—9.

GOODE, J. E., HIGGS, K. H. and HYRYCZ, K. J. (1978), 'Nitrogen and water effects on the nutrition, growth, crop yield and fruit quality of orchard-grown Cox's Orange Pippin apple trees', *Journal of Horticultural Science 53*, 295—306.

GOODE, J. E. HIGGS, K. H. and HYRYCZ, K. J. (1978), 'Trickle irrigation of apple trees and the effects of liquid feeding with NO_3^- and K^+ compared with normal manuring', *Journal of Horticultural Science 53*, 307—16.

GOODE, J. E., HIGGS, K. H. and HYRYCZ, K. J. (1979), 'Effects of water stress control in apple trees by misting', *Journal of Horticultural Science 54*, 1—11.

GOODE, J. E. and HYRYCZ, K. J. (1964), 'The response of Laxton's Superb apple trees to different soil moisture conditions', *Journal of Horticultural Science 39*, 254—76.

GOODE, J. E. and HYRYCZ, K. J. (1968), 'The response of Malling Jewel and Malling Exploit raspberries to different soil moisture conditions and straw mulching', *Journal of Horticultural Science 43*, 215—30.

GOODE, J. E., and INGRAM, J. (1971), 'The effect of irrigation on the growth, cropping and nutrition of Cox's Orange Pippin apple trees', *Journal of Horticultural Science 46*, 195—208.

HAMER, P. J. C. (1980), 'Mist irrigation — the physics of misting and the development of a field system', Seminar on the *Efficiency of localised irrigation as affected by the methods of water applications* (November 1979), Bologna, 151—5.

HAMER, P. J. C. (1987), 'Irrigation of fruit. 3. Irrigation methods', *Irrigation News 11*, 23—8.

HAMER, P. J. C. (1989), 'Simulation of the effects of environmental variables on the water requirements for frost protection by overhead sprinkler irrigation', *Journal of Agricultural Engineering Research 42*, 63—75.

HIGGS, K. H. and JONES, H. G. (1989), 'Water use by strawberry in south-east England', *Journal of Horticultural Science 64*, 167—75.

HUGHES, H. M. (1965), 'Strawberry irrigation experiments on a Brickearth soil', *Journal of Horticultural Science 40*, 285—96.

INGRAM, J. (1981), 'Irrigation of fruit crops', *Irrigation News 1*, 16—19.

INGRAM, J. (1984), 'Frost protection of fruit crops', *Irrigation News 6*, 17-22.

INGRAM, J., ROBINS, D. J. and DOWSE, G. A. (1977), 'Water in fruit production', MAFF/ADAS leaflet.

KONGSRUD, K. L. (1969), 'Effects of soil moisture tension on growth and yield in blackcurrant and apples', *Acta Agriculturae Scandinavica 19*, 245—57.

MACKERRON, D. K. L. (1982), 'Growth and water use in the red raspberry (*Rubus idaeus* L.), 1. Growth and yield under different levels of soil moisture stress', *Journal of Horticultural Science 57*, 295—306.

MAFF (1982), 'Irrigation', *Reference Book 138*.

ROGERS, W. S. and WILSON, D. J. (1957). 'Strawberry cultivation studies. IV. Developments in matted systems', *Journal of Horticultural Science 32*, 99—107.

THOMAS, C. M. S. and CRISP, C. M. (1987), 'Irrigation of fruit. 2. Soft fruit', *Irrigation News 11*, 18—22.

Chapter 10

Irrigation of Grassland

In this chapter, I have considered irrigation of grassland under three headings:

1. The likely yield benefit from irrigation.
2. The economics of irrigating grassland.
3. A recommended irrigation plan.

The Likely Yield Benefit from Irrigation

The growth of grass is reduced in most years as a result of insufficient rainfall, especially in southern and eastern parts of the United Kingdom. In an attempt to quantify this, a series of experiments comparing irrigated and unirrigated grassland was conducted jointly by ADAS and the Agriculture and Food Research Council (AFRC) at a range of sites. Table 10.1 contains some results from the Institute for Grassland and Animal Production (IGAP) at Hurley in Berkshire, where the weather is likely to be representative of much of the south-east of England. From these results it can be seen that the average yield response from irrigation was 2 t/ha dry matter or around 25 per cent.

The average seasonal pattern of production over the thirteen years is shown in Figure 10.1, where it can be seen how unirrigated swards *on average* show lower production during the period June to September. But the incidence of drought is widely variable in both timing and severity. It can be seen in Table 10.1 that drought reduced yield by 40 per cent in 1976, but only by 10 per cent in 1968. In some years the drought may have an effect early in the season, in other years it may be later.

Drought is the dominant cause of annual variation in growth of grass. Figure 10.2 shows the year-to-year variation in production throughout the season for both irrigated and unirrigated grassland. The variation has been expressed mathematically as a 'coefficient of variation'; a low coefficient signifies consistent year-to-year production, but a high coefficient is associated with very variable growth. The high

Table 10.1   Annual yields of dry matter (t/ha) from perennial ryegrass S24 cut eight times at monthly intervals

Year	Irrigated	Unirrigated
1965	10.0	9.3
1966	10.0	9.0
1967	10.7	9.2
1968	10.8	9.7
1969	8.7	5.7
1970	10.3	7.3
1971	9.0	7.7
1972	8.1	6.1
1973	9.9	7.8
1974	10.9	8.5
1975	8.8	5.3
1976	8.6	5.2
1977	10.3	9.2
Mean	9.7	7.7

Source: Institute for Grassland and Animal Production.

variation shown by unirrigated swards necessitates that the stocking rate be limited as insurance against poor growth in a dry year. Conversely, the lower variation and greater certainty of production with irrigation allows for a higher stocking rate.

Experiments have shown that the response of grass to irrigation is linked strongly to the use of nitrogen. Brereton of the Johnstown Castle Research Centre in Ireland showed that the response to irrigation is small in the absence of applied nitrogen, even if the SMD is allowed to rise to very high levels. As the level of applied nitrogen increased, the response to irrigation increases markedly (Figure 10.3).

Doyle at IGAP, Hurley, has taken many of the results from grassland irrigation studies and produced a mathematical model to estimate the likely yield response to irrigation over a range of conditions. The effect of soil type and summer rainfall can be seen in the data from the model presented in Table 10.2, based on a nitrogen application of 600 kg/ha.

In dry years, on the very lightest soils, irrigation may give a response of 5.0 t/ha dry matter, but an average response for a light soil in the eastern counties of England is likely to be between 2 and 3 t/ha or 25–30 per cent yield increase.

The same model can be used to illustrate the effect of nitrogen on irrigation response. This is shown in Table 10.3. The need for high levels of nitrogen application if the potential returns from irrigation are to be realised is clearly shown.

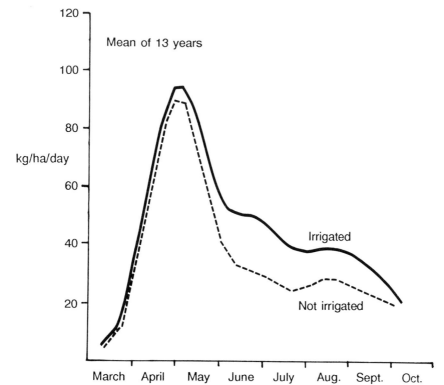

Figure 10.1   Average seasonal pattern of production of irrigated and
unirrigated grassland (perennial ryegrass cv S24).
(Source: Corrall, Institute for Grassland and Animal Production.)

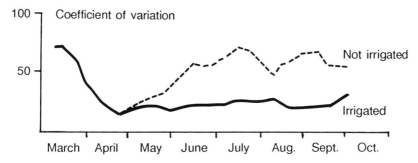

Figure 10.2   Year-to-year variation in the seasonal pattern of grass
production with and without irrigation (perennial ryegrass
cv S24).   (Source: Corrall, Institute for Grassland and Animal
Production.)

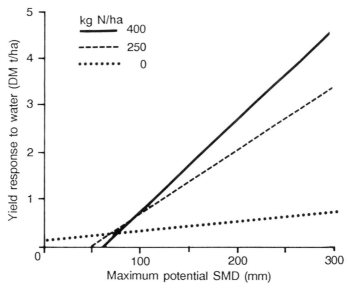

Figure 10.3 *Relationship between grass yield response to irrigation and potential SMD at different levels of nitrogen.*
(Source: Brereton, Johnstown Castle Research Centre.)

Table 10.2   Estimated likely irrigation yield benefits (t/ha DM) under different rainfall regimes and soil AWC

Soil AWC (mm) in top metre	Mean April–September rainfall (mm)				
	280	325	370	415	460
80	5.0	4.2	3.7	3.1	3.1
100	3.2	2.8	2.4	2.0	2.0
130	2.1	1.7	1.6	1.5	1.3

Source: Doyle, Institute for Grassland and Animal Production.

Another benefit of irrigation on grassland was shown by Munro, who demonstrated that irrigation can lead to an improvement in the botanical composition of the sward. This experiment was on a timothy/meadow fescue/white clover ley. At the start of the experiment there was no *Agrostis* sp. and little *Holcus* sp. These increased considerably on plots that were unirrigated, but not on irrigated plots. White clover, on the other hand, increased on the irrigated plots.

Table 10.3 The likely dry matter yield benefits from irrigation at different nitrogen rates (t/ha DM)

Mean summer rainfall (mm)	N kg/ha	Soil AWC (mm)		
		80	100	130
280	150	2.8	1.9	1.2
	300	4.3	2.8	1.9
	450	4.7	3.1	2.1
	600	5.0	3.1	2.1
325	150	2.4	1.6	1.0
	300	3.6	2.4	1.5
	450	4.0	2.6	1.7
	600	4.2	2.8	1.8

Source: Doyle, Institute for Grassland and Animal Production.

THE ECONOMICS OF IRRIGATING GRASSLAND

Investment appraisal for grassland irrigation is complex because the increased growth is not itself usually a directly saleable commodity, but has to be fed to stock to obtain a financial return. If the investment is to be viable, the extra grass produced would probably have to be fed to dairy cows or intensive beef. Stocking rates increase by either carrying more stock or using less land for the original number. Milk quotas may rule out the first option. Where the extra grass produced is conserved to replace compound feed or where land is released for an alternative crop, the returns are often insufficient to warrant investment in irrigation.

As a rule of thumb, irrigation is only likely to be worthwhile in areas where the response averages 2.5 t/ha DM or more. This restricts economic investment to soils with an unusually low AWC of 8 per cent, *almost* regardless of summer rainfall (see Table 10.2) or, more commonly, those with an AWC of 10 per cent if the April–September rainfall is less than 350 mm. In this latter case, the required yield response will only be achieved where fertiliser nitrogen is greater than 300 kg/ha. These calculations assume that water storage facilities are not required, and that the irrigation is applied using mobile rainguns. If a reservoir is needed, or a conventional sprinkler system is used, then it is even more difficult to justify investment.

These are only broad guidelines, however, but serve to identify those farms with the greatest chance of justifying investment in irrigation. Individual farmers contemplating investment in irrigation require an up-to-date financial appraisal specific to their own farming system before making decisions.

Of course, where crops other than grass are already being irrigated, the investment costs associated with irrigating an additional area of pasture may be low enough to justify the investment. On a mixed farm, it is also often possible to irrigate the grass with equipment and water available at a time when the other crops have no requirement for irrigation. This moderately cheap partial irrigation could be a very worthwhile bonus, in particular by insuring against the effects of severe drought in the driest years.

A Recommended Irrigation Plan

The case for intensive irrigation

Some experiments have shown that the critical SMD for grass is relatively low, certainly lower than would be expected from examination of the soil type, available water and rooting depth of the crop. Indeed, some authors have suggested that grass will lose potential yield if the deficit is allowed to reach 25 mm, even on heavy soils.

Garwood and Williams, however, have shown that the crop is fully capable of extracting water from depth, even when deficits are considerably higher than 25 mm, and they have suggested that the crop is not directly sensitive to moisture stress at such deficits. They offer instead an explanation based on nutrient availability, particularly nitrogen. During the growing season available nitrogen is largely concentrated in the upper soil horizons, and the uptake of nitrogen is depressed as the surface layer is depleted of water, despite the ability of the crop to remove further water from depth. The response to irrigation at low soil moisture deficits comes from the fact that water in the surface layers is preferentially absorbed by the crop, along with the nitrogen that is also confined to these upper layers.

Garwood and Williams showed that this indirect effect of drought at low soil moisture deficit could also be overcome by injecting nitrogen into the lower profile where water was still available.

Experiments have shown that, in order to obtain the highest possible yield of grass, the soil should be kept at a very low SMD. In practice, however, farmers are unlikely to irrigate when the SMD is less than 25 mm. Anything less implies irrigating at a great frequency, which is too costly in terms of labour and equipment. It would inevitably result in much wastage of water if rain fell shortly after irrigation, a likely event in a typical British summer. There is also the risk of damage from poaching or vehicle rutting on soils that are kept close to field capacity.

I have calculated a set of critical deficits for grass based on the water availability in the upper 20 cm of the soil, ignoring the rooting characteristics of the crop. Thus there is no distinction made between species, or between leys and permanent pasture. This does not imply that all grass species are equally affected by drought. On the contrary, we know that some species (e.g. tall fescue) can maintain some growth at higher deficits than other species. However, if *maximum* yield is dependent on water being available in the upper layers of soil, as suggested by Garwood and Williams, the critical deficits presented in Table 10.4 should serve for most situations.

Table 10.4   Critical soil moisture deficits for grass according to soil texture

Soil texture	Critical deficit (mm)
Sand	25
Loamy medium sand	25
Loamy fine sand	35
Medium sandy loam	35
Sandy clay or sandy clay loam	35
Clay or silty clay	35
Fine sandy loam	35
Clay loam or silty clay loam	35
Sandy silt loam	40
Silt loam	45
Peats	70

The case for partial irrigation

It has been stated that to obtain the highest possible yield, it is necessary to maintain a small SMD. What tactic should be employed when water is in short supply? Garwood and Tyson promoted the concept of partial irrigation in this situation, which entails irrigating to wet the surface layer only, but not returning the soil to field capacity. The water is applied after cutting or grazing and the application of nitrogen. This strategy has several effects:

1. It reduces the likelihood of wastage from rain falling subsequently.
2. The crop obtains some water (and nitrogen, therefore) from the surface layer, but also utilises some of the water from deeper in the profile.
3. Less irrigation is applied.

4. The technique will not usually allow the crop to achieve its full yield potential, but it will produce the highest possible return for each increment of water applied.

In summary, the intensive irrigation plan described previously is the best strategy for maximum yield but, where irrigation is likely to be limited, this technique of partial irrigation is the best option.

The principles involved in developing a strategy when irrigation is limited for other crops are dealt with in chapter 12.

References and Further Reading

BRERETON, A. J. (1982), 'The effect of water on grassland productivity in Ireland', *Irish Journal of Agricultural Research 21*, 227–48.

CORRALL, A. J. (1978), 'The effect of genotype and water supply on the seasonal pattern of grass production', *Proceedings of the 7th General Meeting of the European Grassland Federation*, Gent, 2.23–2.32.

DOYLE, C. J. (1981), 'Economics of irrigating grassland in the United Kingdom', *Grass and Forage Science 36*, 297–306.

GARWOOD, E. A. (1988), 'Water deficiency and excess in grassland: the implications for grass production and for the efficiency of use of N', *Proceedings of a Colloquium on Nitrogen and Water Use by Grassland*, North Wyke Research Station, Devon, 1987, 24–41.

GARWOOD, E. A. and SINCLAIR, J. (1979), 'Use of water by six grass species: 2. Root distribution and use of soil water', *Journal of Agricultural Science, Cambridge 93*, 23–35.

GARWOOD, E. A. and TYSON, K. C. (1973), 'The response of S24 Perennial Ryegrass swards to irrigation. 1. Effects of partial irrigation on DM yield and on the utilization of applied nitrogen', *Journal of the British Grassland Society 28*, 223–33.

GARWOOD, E. A. TYSON, K. C. and SINCLAIR, J. (1979), 'Use of water by six grass species: 1. Dry matter yields and response to irrigation', *Journal of Agricultural Science, Cambridge 93*, 13–24.

GARWOOD, E. A. and WILLIAMS, T. E. (1967), 'Soil water use and growth of a grass sward', *Journal of Agricultural Science, Cambridge 68*, 281–92.

GARWOOD, E. A. and WILLIAMS, T. E. (1967), 'Growth, water use and nutrient uptake from the subsoil by grass swards', *Journal of Agricultural Science, Cambridge 69*, 125–30.

MUNRO, I. A. (1958), 'Irrigation of grassland', *Journal of the British Grassland Society 13*, 213.

PENMAN, H. L. (1962), 'Woburn irrigation, 1951–9. 2. Results for grass', *Journal of Agricultural Science, Cambridge 58*, 349–64.

STILES, W. and WILLIAMS, T. E. (1965), 'The response of a ryegrass – white clover sward to various irrigation regimes', *Journal of Agricultural Science, Cambridge 65*, 351–64.

WILLIAMS, T. E. (1970), 'Soil water use and irrigation of grass swards', *Annual Report 1969, The Grassland Research Institute*, Hurley, 136–9.

Chapter 11

Water Quality

Various substances can find their way into water supplies and be harmful to crops if the water is used for irrigation. These include boron, calcium, iron, sodium salts and some organic contaminants. The element which most frequently limits the suitability of water for irrigation is chloride, usually derived from dissolved sodium chloride (common salt).

SALINE WATER

Saline water contains both sodium and chloride. Problems are most commonly encountered in coastal areas where sea water finds its way into ground water or surface supplies. Some rock strata naturally contain high levels of salt, and borehole water can be saline to varying degrees. Water pumped from the salt-bearing strata into surface rivers, as happens in some coal mining areas, can lead to salt levels in the river which are too high for irrigation purposes. Although the sodium can cause damage to soil structure in clay soils, it is the problems associated with the chloride content of saline water which are of greater concern. Chloride can damage plants in two ways: by directly scorching the leaves and by inhibition of root activity.

Direct leaf scorch is usually associated with bright, sunny conditions, and if saline water has to be used the risk can be reduced by irrigating during dull periods or in the evening. It is difficult to be precise about the levels of chloride associated with such damage, because even an initially low concentration of chloride in irrigation water can soon increase to a high concentration as the droplets on the foliage evaporate and decrease in size.

The effect of chloride on root activity is related to the concentration of chloride in the soil solution. Water containing 500 mg/litre chloride could be used for a single application of 25 mm on the potato crop without undue effect. However, further applications would increase the accumulation of chloride in the soil and so increase the risk. Heavier soils, with their higher AWC, can receive greater amounts of chloride before the concentration reaches harmful levels.

Safe guidelines for maximum chloride content are given in Table 11.1. Crops are listed according to their sensitivity, along with the maximum safe levels of chloride that can be applied to each.

Table 11.1 Crop sensitivity to chloride, and maximum amounts of irrigation related to chloride concentration

Tolerance groups	Crops		Chloride in water (mg/litre)						
			100	200	300	400	500	600	Over 600
			Maximum total irrigation (mm)						
Very sensitive	Peas Dwarf beans	Strawberries Blackberries Gooseberries Plums	65	30	NS	NS	NS	NS	NS
Moderately sensitive	Beans Lettuce Onion Celery Maize Clover	Apples Pears Raspberries Redcurrants	140	70	45	NS	NS	NS	NS
Slightly sensitive	Potatoes Cabbage Carrots Cauliflower Wheat Ryegrass	Blackcurrants Vines	250+	130	90	60	30	NS	NS
Least sensitive	Sugar beet Red beet Spinach Kale Barley		250+	250+	160	120	100	75	NS

Source: ADAS.
Note: NS = Not suitable.
    The figures above are for a loamy sand with an available water capacity of about 120 mm per metre depth. Greater concentrations are acceptable on soils with a higher AWC. The above values can be increased 50 per cent where the AWC is 180 mm per metre depth (e.g. a fine sandy loam).

The recommendations in this table are designed to minimise the risk of damage by chloride, but in some cases the use of water containing higher chloride levels can be considered.

As an example, consider two experiments conducted at Gleadthorpe

EHF in 1979 and 1983. Potatoes were either left unirrigated or irrigated with water containing different concentrations of chloride. The results are shown in Table 11.2.

Table 11.2 Effect of chloride concentration in irrigation water on yield of potatoes

Chloride concentration in irrigation water (mg/litre)	Ware yield (t/ha)	
	1979	1983
No irrigation	33.8	23.4
0	38.6	35.0
500	44.8	31.7
1000	41.1	30.6
1500	37.0	29.4
2000	40.3	26.8

Source: ADAS.

It can be seen that increasing levels of chloride in the water decreased yield compared with using clean water without any chloride, but it is also apparent that the use of water containing 2,000 mg/litre chloride resulted in a better yield than leaving the crop unirrigated.

When the levels of chloride concentration exceed those in Table 11.1, what action should a grower take?

If the crop is a vegetable and the leaf is the major edible portion, it is obvious that the guidelines must be adhered to and high chloride water should not be used. Any foliar scorch could render the whole crop virtually unmarketable.

With other crops, it will be possible to exceed these guidelines somewhat, but it is difficult to advise precisely. I would irrigate potatoes with water containing 1,000 mg/litre chloride, rather than risk any drought effects. At 1,500 mg/litre, it is still probably better to irrigate, but there is an unpredictable risk of scorch. Water containing 2,000 mg/litre chloride should only be used if the situation is desperate, because the risk of scorch is quite high.

If more than 75 mm irrigation has been applied using water with more than 1,200 mg/litre chloride, a grower would be advised to have his soil examined, including a conductivity assessment. This will provide a check on any potential problems arising from the accumulation of chloride in the soil. Where the accumulation of chloride in the soil solution significantly exceeds that in the irrigation water, it may be possible to irrigate excessively and leach much of the accumulated chloride to lower layers. Expert advice should be sought before attempting to do this.

If a grower is uncertain of the chloride levels in his irrigation water, he should purchase a salinity testing kit. Such kits are particularly useful for variable river water, as found in some coal mining areas, or for sources near the sea. For boreholes, an occasional laboratory test would be adequate.

BORON

Although boron is an essential plant nutrient, it can be damaging if excessive amounts are applied. Boron is commonly used as a bleach in household detergents as well as in industrial processes, so river water can sometimes contain high levels, as a result of contamination by sewage or industrial effluent. Unpolluted waters seldom contain more than 0.5 mg/litre of boron.

The safe levels for boron application are considered to be 2, 3 and 4 kg/ha annually for sensitive, intermediate and tolerant crops respectively. Thus the safe level within irrigation water will depend on the amount of water to be applied, as with chloride levels discussed in the previous section. These safe levels are shown in Table 11.3. It is usual for protected crops to receive around 500 mm of water annually, and the boron concentration should not then exceed 0.4 mg/litre with sensitive crops, 0.6 mg/litre with intermediate crops and 0.8 mg/litre with tolerant crops.

OTHER ELEMENTS

Calcium is frequently found in high levels in water supplies in the south and east of the country, originating from natural chalk and limestone sources. Crops do not suffer any direct toxic effects from this, and often benefit from the increased pH associated with such water. Some pot plants and container-grown nursery stock require acid conditions, however, and the pH must then be neutralised by the use of ammonium salts, or nitric or phosphoric acids.

Use of calcium-rich irrigation water can result in a white deposit on the leaves and fruits, and blockage of trickle irrigation systems. The removal of calcium from the water is too expensive to contemplate for irrigation. A cheaper method, softening the water with sodium, is likely to make the water unsuitable for irrigation.

Table 11.3   Safe boron concentrations in irrigation water according to
                     seasonal water need

Tolerance group	Crops		Seasonal irrigation need (mm)		
			50	100	200
			Safe boron concentration (mg/litre)		
Sensitive	Plums	Blackcurrants			
	Pears	Strawberries			
	Apples	Raspberries			
	Cherries		4.0	2.0	1.0
Intermediate	Barley	Potatoes			
	Wheat	Peas			
	Maize	Radish			
	Oats	Tomatoes	6.0	3.0	1.5
Tolerant	Asparagus	Onions			
	Beet	Cabbage			
	Mangolds	Lettuce			
	Lucerne	Carrots			
	Broad beans	Turnips	8.0	4.0	2.0

Source: ADAS.

Sulphur, in the form of sulphate, can also result in a white deposit on fruit and leaves, as well as blockage of equipment, but is unlikely to have any direct deleterious effect on crop growth.

Iron is sometimes found in underground sources of water, and can cause brown leaf spotting on sensitive crops such as lettuce, as well as blockage of trickle systems. For lettuce and fruit crops where spotting is a problem, the level should not exceed 1 mg/litre. Aeration of the water, followed by a period of settling, can be used to remove iron, so reservoir storage can alleviate the problem, especially if the water is aerated at the point of entry.

Nitrates. Most water supplies contain some nitrates but, while they may be considered unsuitable for drinking purposes, they are not usually damaging to irrigated crops. There are exceptions, such as recirculated water in glasshouses, where levels can reach as high as 200 mg/litre, which may be damaging to a number of exotic species. In this situation, frequent analysis of the water is advisable to avoid the risk of crop damage.

Fluoride levels are usually quite low (below 1 mg/litre) and not damaging to crops. However, there is some Dutch research that has shown that gladioli, freesias and some other flower crops can be damaged by even low levels of fluoride. Work has also shown that the vase life of some cut flowers can be reduced by fluoride at low concentrations.

Trace elements.   Some of these can cause damage if applied to excess, and ADAS has published guidelines on the limits to be used (see Table 11.4). These guidelines apply to protected crops receiving 500 mm water annually, over a period of fifty years. They could also be applied to outdoor crops which receive much less irrigation, with the certain knowledge that a good safety margin is included.

Table 11.4   Suggested maximum trace element concentrations for irrigation water

Trace element	Concentration (mg/litre)
Arsenic	0.04
Cadmium	0.02
Chromium	2.00
Copper	0.50
Molybdenum	0.03
Nickel	0.15
Selenium	0.02
Zinc	1.00
Lead	2.00

Source: ADAS.

Some agrochemicals could cause problems if they found their way into irrigation water, even at very low concentrations. Herbicides, in particular, have caused problems in the past.

Some micro-organisms can also cause problems. Many vegetable crops are eaten without cooking, and the need to avoid the use of contaminated water is obvious.

Further, there is a real risk that some plant pathogens, for example the causal organism of *Rhizomania*, could be spread in irrigation water.

Finally, even the presence of sand or silt in water can render it unsuitable for irrigation, because high levels can cause undue wear on pumps and contribute to blockages in equipment.

References and Further Reading

NEEDHAM, P. (1985), 'Irrigation water quality. 1. Criteria', *Irrigation News 8*, 47–53.

WILLIAMS, J. H. (1972), 'Water quality criteria for crop irrigation (chloride, boron and sodium)', *ADAS Quarterly Review 7*, 106–22.

WILLIAMS, J. H. (1981), 'Water quality for crop irrigation: guidelines on chemical criteria', *ADAS leaflet 776.*

Chapter 12

Limited Water Availability

Choosing Which Crops to Irrigate

Sooner or later, most growers with irrigation facilities are forced to make a decision as to which crops should be given priority. For some the problem may occur only in an exceptionally dry year but, for most farmers, it is a common situation occurring nearly every year. Difficulties may arise from having insufficient water or too little application equipment. Whatever the reason, it is important to understand the relative value of irrigating different crops at various stages through the season.

Although there are many published accounts showing a range of irrigation responses for each crop, there have been very few experiments comparing the response of different crops under identical conditions. The work at Gleadthorpe EHF is interesting in this respect because, since 1985, several crops have been grown together in one field under a specially adapted linear move irrigator (see Colour plate 2) and their irrigation responses have been recorded. It is valid to compare responses from different crops in this study, not only because they have been measured under identical conditions and broadly similar soil, but also because any bias due to slight variation in soil conditions has been reduced as much as possible by rotating the crops around each year. The yield responses and associated benefits are shown in Table 12.1.

Table 12.1   Average yield responses to irrigation at Gleadthorpe EHF, 1985–8

Crop	Yield response (t/ha mm)	Assumed price (£/tonne)	Financial benefits (£/ha mm)
Carrots	0.13	80	10.40
Potatoes	0.07	70	4.90
Sugar beet	0.03	30	0.90
Onions	0.006	95	0.57
Cereals	0.0006	100	0.06

Source: Gleadthorpe EHF, ADAS.

If we are to compare the true value of irrigation on each of these crops, the quality factor must be considered. Irrigation improves the quality of potatoes more than the other four crops listed in Table 12.1. It is difficult to value this, but the improved quality of irrigated potatoes in a dry year is likely to increase the financial benefits obtained from irrigating potatoes above those obtained from irrigating carrots. This also applies to some other vegetable crops where high quality produce is essential and obtainable only with irrigation.

This exercise demonstrates an obvious but important point. The economic returns from irrigating crops can vary tremendously according to the crop i.e. more than a hundred-fold. It may be tempting to irrigate a large area of the farm in a drought situation, but if irrigating cereals is likely to risk the supply of water available to potatoes or carrots later in the season, it should not even be considered. It is important that each grower is aware of the likely irrigation requirements of each of his high-return crops in a dry year, and not put these at risk in the attempt to improve yield of low-return crops on the farm.

Unfortunately, there have not been enough of these comparisons to draw up a scientifically based league table of priorities for a wide range of crops. However, in an attempt to provide some broad guidelines, I have examined the available data already described in various sections of this book and tentatively produced a classification of the financial benefit associated with the extra yield that can be expected from irrigation. This classification is presented in Table 12.2 but, as already stated, this is only part of the story.

In many cases, irrigation will not only produce extra yield, but will affect quality and increase the value of *the entire crop*. Consider apples as an example. Although the yield increase and associated financial returns are relatively low, there is a considerable financial benefit from attaining a fruit above 65 mm diameter, and without skin-cracking. As described in chapter 9, irrigation increases the likelihood of achieving these objectives. For this reason, apples will justly occupy a high position in a list of priorities. I have marked the obvious similar situations in Table 12.2, but there are others depending on individual circumstances, e.g. the market for which the crop is destined. Each grower will need to make some adjustment to the table to produce a priority rating for his own situation. For instance, growers intending to market potatoes with high quality skin-finish may place a higher priority on this crop during the common scab control phase, but those growing potatoes for processing may not. Another example is provided by lettuces. A grower is likely to place a very high priority on these in the weeks leading up to harvest, but possibly a lower priority in the earlier part of the season.

Table 12.2   Average yield response obtained from 25 mm irrigation and its financial value

Crop	Yield response from 25 mm irrigation (t/ha)	Value of crop (£/tonne) at time of writing	Value of extra yield
Green beans (fresh	1.5	725	1088
Raspberries (fresh)	0.625	1350	844*
Strawberries (fresh)	0.625	1100	688*
Lettuce (crisp)	250 boxes	2.65 /box	663*
Cauliflowers	250 crates	2.50 /crate	625
Runner beans	1.25	450	563
Cabbage	3.5	120	420
Strawberries (processing)	0.625	585	366
Lettuce (round)	250 boxes	1.45 /box	363
Brussel sprouts	1.0	275	275
Carrots	3.25	80	260
Green beans (processing)	1.5	170	255
Blackcurrants (processing)	0.75	325	244
Vining peas	1.0	170	170
Early potatoes	1.25	130	163
Maincrop potatoes	2.25	70	158*
Onions	1.25	95	119
Apples	0.375	200	75*
Sugar beet	2.25	29	65
Grassland	0.625 (dry matter)	38	24
Cereals	0.175	110	19

Source: MAFF, ADAS.
Note: * Indicates that irrigation has significant effect on quality, and true value is much greater than indicated by this analysis.

PARTIAL IRRIGATION OF A LARGE AREA OR FULL IRRIGATION OF A SMALL AREA?

As already described, it is a relatively easy decision to choose between crops when the likely returns are of markedly different value. It is much more complex when there are several crops likely to show a similar return, or even a large area of a single crop that cannot be irrigated sufficiently. A choice must then be made between irrigating one crop, or even part of a crop, adequately, or applying a smaller amount of water over the whole area.

This choice hinges entirely on the nature of the crop concerned, and the market for which it is intended. Before discussing this choice in detail it is important to understand the general relationship between crop yield and water. This was demonstrated in Figure 2.1 (in chapter 2), which shows a slight loss of yield associated with the initial stages of soil water depletion, but an increasingly steeper yield loss as the depletion of soil water becomes more pronounced.

This can be expressed in another way. We are all familiar with graphs showing the response of crops to fertiliser. There is a steady rate of increase until the fertiliser level exceeds a certain value, when the rate of increase in yield declines. Experiments have shown that crops respond to water in a similar way (see Figures 12.1 and 12.2), and yield response is subject to the law of diminishing returns.

With a crop such as sugar beet, where yield is the major objective, this determines our strategy. The initial applications are more efficient in promoting yield than are subsequent applications, and a small

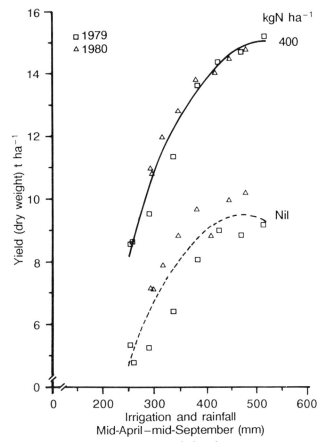

Figure 12.1   Response of grass to irrigation.   (Source: Carr, Silsoe College.)

amount of irrigation spread evenly over the whole crop is the surest
way of achieving the highest yield overall. Further, there is always
the possibility of some rain in a British summer, and this again will
be utilised at maximum efficiency if the irrigation has been spread
over the crop, rather than concentrated in a smaller area.

Other crops best treated in this manner include grass, cereals and
any situation in which the major objective is to obtain the greatest overall
yield.

Conversely, if quality improvement is the major objective in irrigation, as is often the case with potatoes and some field vegetables,
any moisture stress could result in poor quality (scab, growth cracks,
etc.) causing the whole crop to be downgraded. In this situation it would
be unwise to spread irrigation thinly, as it may result in a high yield

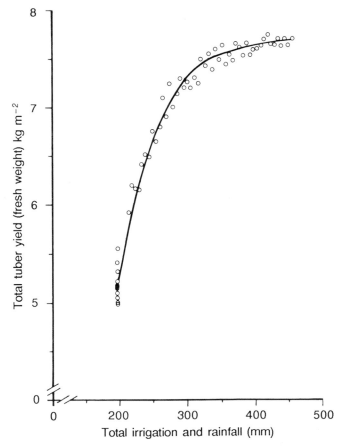

Figure 12.2   Response of potatoes (cv Record) to irrigation.
(Source: Carr, Silsoe College.)

of poor quality produce. Under such circumstances, it may be safer to concentrate on irrigating a smaller area properly, at least guaranteeing the desired quality for part of the crop.

Almost every situation is different, and choosing the best strategy is a skilled management exercise. As an example, consider a farm with several varieties of potatoes. For many years, ADAS has collected data to show how different varieties respond to irrigation (see Table 12.3).

As explained in chapter 5, irrigation can also be used to control the disease of common scab on potatoes. Knowledge of a particular variety's yield response to irrigation and common scab resistance rating, along with the expected value of the crop, can be used to design an irrigation strategy when resources are limited. Estima is quite susceptible to common scab, Pentland Crown is resistant. Record is produced for processing and is therefore saleable with moderate levels of the disease. On a farm growing these varieties, irrigation in the early part of the season should concentrate on controlling common scab in the variety Estima. After the scab control period (see chapter 5), yield response becomes the major consideration. Table 12.3 shows that Estima and Record show a higher response to irrigation than does Pentland Crown. The best strategy would, therefore, involve irrigating Estima and Record, leaving Pentland Crown unirrigated until such time

Table 12.3 Estimates of irrigation response at Gleadthorpe EHF and NIAB (National Institute of Agricultural Botany) common scab resistance rating for different varieties

Variety	Average yield increase (t/ha)	Resistance to common scab*
Estima	14	5
Record	13	6
Wilja	12	6
Cara	12	6
King Edward	12	6
Maris Piper	11	2
Romano	11	5
Desirée	11	3
Pentland Crown	10	8
Kingston	9	7
Pentland Dell	9	5
Pentland Squire	7	6

Source: Gleadthorpe EHF, ADAS and NIAB.
Note: * Resistance to common scab: 1 = susceptible, 9 = resistant.

as spare irrigation capacity is available. If, however, the varieties included Pentland Squire for 'baker' production, the importance of obtaining a bold sample would require that this variety be irrigated intensively, even though its average yield response is lower than the other varieties. There is a need to balance the value of the average yield responses against the risk of a whole crop being downgraded.

In many cases, crops require very intensive irrigation if they are to achieve their maximum potential yield but, where water is limited, a single application of 25 mm at a certain stage of growth may be sufficient to produce fairly high yields of the required quality. Examples include early summer cauliflowers irrigated when the curds reach a diameter of 30 mm, or cabbages during the last three weeks before harvest. Precise knowledge of the responsive stages of each crop will enable growers to get the best returns from a limited quantity of water and equipment.

The management of irrigated crops commercially involves such decisions. It is hoped that by describing in this book some of the results of the large number of experiments that have been conducted over the years, along with some of my personal views on the subject, I have been of assistance to farmers, growers and advisers facing these complex decisions.

REFERENCES AND FURTHER READING

CARR, M. K. V. (1983), 'Identifying the need for irrigation', *Irrigation News 5*, 21–9.
CARR, M. K. V. (1985), 'The role of water in the productivity of potatoes', in M. K. V. Carr and P. J. C. Hamer (eds), 'Irrigating Potatoes', *UK Irrigation Association Technical Monograph 2*, 1–11.
MAFF (1982), 'Irrigation', *Reference Book 138*.

Index

187

FARMING PRESS BOOKS

Below is a sample of the wide range of agricultural books published by Farming Press. For more information or a free illustrated book list please contact:

**Farming Press Books, 4 Friars Courtyard
30–32 Princes Street, Ipswich IP1 1RJ, United Kingdom
Telephone (0473) 43011**

Cereal Pests and Diseases
R. Gair, J. E. E. Jenkins and E. Lister
An outstanding guide to the recognition and control of cereal pests and diseases by two of Britain's foremost plant pathologists and a leading entomologist.

Drying and Storing Combinable Crops
K. A. McLean
The standard reference, fully revised in an enlarged second edition.

Cereal Husbandry
John Wibberley
A wide-ranging exposition of the principles of temperate cereal production.

Crop Nutrition and Fertiliser Use
John Archer
Gives details of uptake for each nutrient and then deals with the specific requirements of temperate crops.

Oilseed Rape
J. T. Ward, W. D. Basford, J. H. Hawkins and J. M. Holliday
Contains up-to-date information on all aspects of oilseed rape growth, nutrition, pest control and marketing.

Farm Machinery
Brian Bell
Gives a sound introduction to a wide range of tractors and farm equipment. Now revised, enlarged and incorporating over 150 photographs.

Soil Management
D. B. Davies, D. J. Eagle and B. Finney
Two soil scientists and a senior mechanisation officer with ADAS go into all aspects of the soil, plant nutrition, farm implements and the effects on the soil, crop performance, land drainage and cultivation system.

Farming Press also publish four monthly magazines: *Arable Farming*, *Livestock Farming*, *Pig Farming* and *Dairy Farmer*. For a specimen copy of any of these magazines please contact Farming Press at the address above.